FIRST COURSE

IN THE

THEORY OF EQUATIONS

BY

LEONARD EUGENE DICKSON, Ph.D.

CORRESPONDANT DE L'INSTITUT DE FRANCE
PROFESSOR OF MATHEMATICS IN THE UNIVERSITY OF CHICAGO

Merchant Books

1922

PREFACE

The theory of equations is not only a necessity in the subsequent mathematical courses and their applications, but furnishes an illuminating sequel to geometry, algebra and analytic geometry. Moreover, it develops anew and in greater detail various fundamental ideas of calculus for the simple, but important, case of polynomials. The theory of equations therefore affords a useful supplement to differential calculus whether taken subsequently or simultaneously.

It was to meet the numerous needs of the student in regard to his earlier and future mathematical courses that the present book was planned with great care and after wide consultation. It differs essentially from the author's *Elementary Theory of Equations*, both in regard to omissions and additions, and since it is addressed to younger students and may be used parallel with a course in differential calculus. Simpler and more detailed proofs are now employed. The exercises are simpler, more numerous, of greater variety, and involve more practical applications.

This book throws important light on various elementary topics. For example, an alert student of geometry who has learned how to bisect any angle is apt to ask if every angle can be trisected with ruler and compasses and if not, why not. After learning how to construct regular polygons of 3, 4, 5, 6, 8 and 10 sides, he will be inquisitive about the missing ones of 7 and 9 sides. The teacher will be in a comfortable position if he knows the facts and what is involved in the simplest discussion to date of these questions, as given in Chapter III. Other chapters throw needed light on various topics of algebra. In particular, the theory of graphs is presented in Chapter V in a more scientific and practical manner than was possible in algebra and analytic geometry.

There is developed a method of computing a real root of an equation with minimum labor and with certainty as to the accuracy of all the decimals obtained. We first find by Horner's method successive transformed equations whose number is half of the desired number of significant figures of the root. The final equation is reduced to a linear equation by applying to the constant term the correction computed from the omitted terms of the second and higher degrees, and the work is completed by abridged division. The method combines speed with control of accuracy.

Newton's method, which is presented from both the graphical and the numerical standpoints, has the advantage of being applicable also to equations which are not algebraic; it is applied in detail to various such equations.

In order to locate or isolate the real roots of an equation we may employ a graph, provided it be constructed scientifically, or the theorems of Descartes, Sturm, and Budan, which are usually neither stated, nor proved, correctly.

The long chapter on determinants is independent of the earlier chapters. The theory of a general system of linear equations is here presented also from the standpoint of matrices.

For valuable suggestions made after reading the preliminary manuscript of this book, the author is greatly indebted to Professor Bussey of the University of Minnesota, Professor Roever of Washington University, Professor Kempner of the University of Illinois, and Professor Young of the University of Chicago. The revised manuscript was much improved after it was read critically by Professor Curtiss of Northwestern University. The author's thanks are due also to Professor Dresden of the University of Wisconsin for various useful suggestions on the proof-sheets.

CHICAGO, 1921.

CONTENTS

Numbers refer to pages.

v

CHAPTER IV
Cubic and Quartic Equations

CHAPTER V
The Graph of an Equation

CHAPTER VI
Isolation of Real Roots

CHAPTER VII
Solution of Numerical Equations

CHAPTER VIII
Determinants; Systems of Linear Equations

DIGITALIZED BY
WATCHMAKER PUBLISHING
ALL RIGHTS RESERVED

First Course in
The Theory of Equations

———

CHAPTER I

Complex Numbers

1. Square Roots. If p is a positive real number, the symbol $\sqrt{p}$ is used to denote the positive square root of p. It is most easily computed by logarithms.

We shall express the square roots of negative numbers in terms of the symbol i such that the relation $i^2 = -1$ holds. Consequently we denote the roots of $x^2 = -1$ by i and $-i$. The roots of $x^2 = -4$ are written in the form $\pm 2i$ in preference to $\pm\sqrt{-4}$. In general, if p is positive, the roots of $x^2 = -p$ are written in the form $\pm\sqrt{p}\,i$ in preference to $\pm\sqrt{-p}$.

The square of either root is thus $(\sqrt{p})^2 i^2 = -p$. Had we used the less desirable notation $\pm\sqrt{-p}$ for the roots of $x^2 = -p$, we might be tempted to find the square of either root by multiplying together the values under the radical sign and conclude erroneously that

$$\sqrt{-p}\sqrt{-p} = \sqrt{p^2} = +p.$$

To prevent such errors we use $\sqrt{p}\,i$ and not $\sqrt{-p}$.

2. Complex Numbers. If a and b are any two real numbers and $i^2 = -1$, $a + bi$ is called a *complex number*[1] and $a - bi$ its *conjugate*. Either is said to be *zero* if $a = b = 0$. Two complex numbers $a + bi$ and $c + di$ are said to be *equal* if and only if $a = c$ and $b = d$. In particular, $a + bi = 0$ if and only if $a = b = 0$. If $b \neq 0$, $a + bi$ is said to be *imaginary*. In particular, bi is called a *pure imaginary*.

---

[1] Complex numbers are essentially couples of real numbers. For a treatment from this standpoint and a treatment based upon vectors, see the author's *Elementary Theory of Equations*, p. 21, p. 18.

Addition of complex numbers is defined by

$$(a + bi) + (c + di) = (a + c) + (b + d)i.$$

The inverse operation to addition is called subtraction, and consists in finding a complex number z such that

$$(c + di) + z = a + bi.$$

In notation and value, z is

$$(a + bi) - (c + di) = (a - c) + (b - d)i.$$

Multiplication is defined by

$$(a + bi)(c + di) = ac - bd + (ad + bc)i,$$

and hence is performed as in formal algebra with a subsequent reduction by means of $i^2 = -1$. For example,

$$(a + bi)(a - bi) = a^2 - b^2 i^2 = a^2 + b^2.$$

Division is defined as the operation which is inverse to multiplication, and consists in finding a complex number q such that $(a+bi)q = e+fi$. Multiplying each member by $a - bi$, we find that q is, in notation and value,

$$\frac{e + fi}{a + bi} = \frac{(e + fi)(a - bi)}{a^2 + b^2} = \frac{ae + bf}{a^2 + b^2} + \frac{af - be}{a^2 + b^2}i.$$

Since $a^2 + b^2 = 0$ implies $a = b = 0$ when a and b are real, we conclude that division except by zero is possible and unique.

EXERCISES

Express as complex numbers

1. $\sqrt{-9}$.   **2.** $\sqrt{4}$.

3. $(\sqrt{25} + \sqrt{-25})\sqrt{-16}$.   **4.** $-\frac{2}{3}$.

5. $8 + 2\sqrt{3}$.   **6.** $\dfrac{3 + \sqrt{-5}}{2 + \sqrt{-1}}$.   **7.** $\dfrac{3 + 5i}{2 - 3i}$.   **8.** $\dfrac{a + bi}{a - bi}$.

9. Prove that the sum of two conjugate complex numbers is real and that their difference is a pure imaginary.

10. Prove that the conjugate of the sum of two complex numbers is equal to the sum of their conjugates. Does the result hold true if each word sum is replaced by the word difference?

11. Prove that the conjugate of the product (or quotient) of two complex numbers is equal to the product (or quotient) of their conjugates.

12. Prove that, if the product of two complex numbers is zero, at least one of them is zero.

13. Find two pairs of real numbers x, y for which

$$(x + yi)^2 = -7 + 24i.$$

As in Ex. 13, express as complex numbers the square roots of

14. $-11 + 60i.$ **15.** $5 - 12i.$ **16.** $4cd + (2c^2 - 2d^2)i.$

3. Cube Roots of Unity. Any complex number x whose cube is equal to unity is called a *cube root of unity*. Since

$$x^3 - 1 = (x - 1)(x^2 + x + 1),$$

the roots of $x^3 = 1$ are 1 and the two numbers x for which

$$x^2 + x + 1 = 0, \qquad (x + \tfrac{1}{2})^2 = -\tfrac{3}{4}, \qquad x + \tfrac{1}{2} = \pm\tfrac{1}{2}\sqrt{3}i.$$

Hence there are three cube roots of unity, viz.,

$$1, \qquad \omega = -\tfrac{1}{2} + \tfrac{1}{2}\sqrt{3}i, \qquad \omega' = -\tfrac{1}{2} - \tfrac{1}{2}\sqrt{3}i.$$

In view of the origin of ω, we have the important relations

$$\omega^2 + \omega + 1 = 0, \quad \omega^3 = 1.$$

Since $\omega\omega' = 1$ and $\omega^3 = 1$, it follows that $\omega' = \omega^2$, $\omega = \omega'^2$.

4. Geometrical Representation of Complex Numbers. Using rectangular axes of coördinates, OX and OY, we represent the complex number $a + bi$ by the point A having the coördinates a, b (Fig. 1).

The positive number $r = \sqrt{a^2 + b^2}$ giving the length of OA is called the *modulus* (or *absolute value*) of $a+bi$. The angle $\theta = XOA$, measured counter-clockwise from OX to OA, is called the *amplitude* (or *argument*) of $a + bi$. Thus $\cos\theta = a/r$, $\sin\theta = b/r$, whence

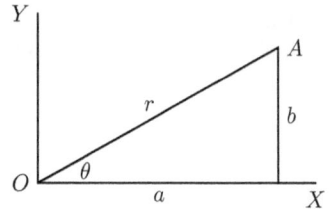

FIG. 1

(1) $$a + bi = r(\cos\theta + i\sin\theta).$$

The second member is called the *trigonometric form* of $a + bi$.

For the amplitude we may select, instead of θ, any of the angles $\theta \pm 360°$, $\theta \pm 720°$, etc.

Two complex numbers are equal if and only if their moduli are equal and an amplitude of the one is equal to an amplitude of the other.

For example, the cube roots of unity are 1 and

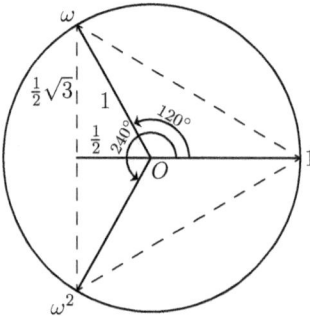

FIG. 2

$$\omega = -\tfrac{1}{2} + \tfrac{1}{2}\sqrt{3}i$$
$$= \cos 120° + i\sin 120°,$$
$$\omega^2 = -\tfrac{1}{2} - \tfrac{1}{2}\sqrt{3}i$$
$$= \cos 240° + i\sin 240°,$$

and are represented by the points marked 1, ω, ω^2 at the vertices of an equilateral triangle inscribed in a circle of radius unity and center at the origin O (Fig. 2). The indicated amplitudes of ω and ω^2 are 120° and 240° respectively, while the modulus of each is 1.

The modulus of -3 is 3 and its amplitude is 180° or 180° plus or minus the product of 360° by any positive whole number.

5. Product of Complex Numbers. By actual multiplication,

$$\left[r(\cos\theta + i\sin\theta)\right]\left[r'(\cos\alpha + i\sin\alpha)\right]$$
$$= rr'\left[(\cos\theta\cos\alpha - \sin\theta\sin\alpha) + i(\sin\theta\cos\alpha + \cos\theta\sin\alpha)\right]$$
$$= rr'\left[\cos(\theta + \alpha) + i\sin(\theta + \alpha)\right], \quad \text{by trigonometry.}$$

Hence *the modulus of the product of two complex numbers is equal to the product of their moduli, while the amplitude of the product is equal to the sum of their amplitudes.*

For example, the square of $\omega = \cos 120° + i\sin 120°$ has the modulus 1 and the amplitude $120° + 120°$ and hence is $\omega^2 = \cos 240° + i\sin 240°$. Again, the product of ω and ω^2 has the modulus 1 and the amplitude $120° + 240°$ and hence is $\cos 360° + i \sin 360°$, which reduces to 1. This agrees with the known fact that $\omega^3 = 1$.

Taking $r = r' = 1$ in the above relation, we obtain the useful formula

$$(2) \qquad (\cos \theta + i \sin \theta)(\cos \alpha + i \sin \alpha) = \cos(\theta + \alpha) + i \sin(\theta + \alpha).$$

6. Quotient of Complex Numbers. Taking $\alpha = \beta - \theta$ in (2) and dividing the members of the resulting equation by $\cos \theta + i \sin \theta$, we get

$$\frac{\cos \beta + i \sin \beta}{\cos \theta + i \sin \theta} = \cos(\beta - \theta) + i \sin(\beta - \theta).$$

Hence *the amplitude of the quotient of $R(\cos \beta + i \sin \beta)$ by $r(\cos \theta + i \sin \theta)$ is equal to the difference $\beta - \theta$ of their amplitudes, while the modulus of the quotient is equal to the quotient R/r of their moduli.*

The case $\beta = 0$ gives the useful formula

$$\frac{1}{\cos \theta + i \sin \theta} = \cos \theta - i \sin \theta.$$

7. De Moivre's Theorem. *If n is any positive whole number,*

$$(3) \qquad (\cos \theta + i \sin \theta)^n = \cos n\theta + i \sin n\theta.$$

This relation is evidently true when $n = 1$, and when $n = 2$ it follows from formula (2) with $\alpha = \theta$. To proceed by mathematical induction, suppose that our relation has been established for the values $1, 2, \ldots, m$ of n. We can then prove that it holds also for the next value $m + 1$ of n. For, by hypothesis, we have

$$(\cos \theta + i \sin \theta)^m = \cos m\theta + i \sin m\theta.$$

Multiply each member by $\cos \theta + i \sin \theta$, and for the product on the right substitute its value from (2) with $\alpha = m\theta$. Thus

$$(\cos \theta + i \sin \theta)^{m+1} = (\cos \theta + i \sin \theta)(\cos m\theta + i \sin m\theta),$$
$$= \cos(\theta + m\theta) + i \sin(\theta + m\theta),$$

which proves (3) when $n = m + 1$. Hence the induction is complete.

Examples are furnished by the results at the end of §5:

$$(\cos 120° + i \sin 120°)^2 = \cos 240° + i \sin 240°,$$
$$(\cos 120° + i \sin 120°)^3 = \cos 360° + i \sin 360°.$$

8. Cube Roots. To find the cube roots of a complex number, we first express the number in its trigonometric form. For example,

$$4\sqrt{2} + 4\sqrt{2}i = 8(\cos 45° + i \sin 45°).$$

If it has a cube root which is a complex number, the latter is expressible in the trigonometric form

(4) $r(\cos \theta + i \sin \theta)$.

The cube of the latter, which is found by means of (3), must be equal to the proposed number, so that

$$r^3(\cos 3\theta + i \sin 3\theta) = 8(\cos 45° + i \sin 45°).$$

The moduli r^3 and 8 must be equal, so that the positive real number r is equal to 2. Furthermore, 3θ and $45°$ have equal cosines and equal sines, and hence differ by an integral multiple of $360°$. Hence $3\theta = 45° + k \cdot 360°$, or $\theta = 15° + k \cdot 120°$, where k is an integer.[2] Substituting this value of θ and the value 2 of r in (4), we get the desired cube roots. The values 0, 1, 2 of k give the distinct results

$$R_1 = 2(\cos 15° + i \sin 15°),$$
$$R_2 = 2(\cos 135° + i \sin 135°),$$
$$R_3 = 2(\cos 255° + i \sin 255°).$$

Each new integral value of k leads to a result which is equal to R_1, R_2 or R_3. In fact, from $k = 3$ we obtain R_1, from $k = 4$ we obtain R_2, from $k = 5$ we obtain R_3, from $k = 6$ we obtain R_1 again, and so on periodically.

EXERCISES

1. Verify that $R_2 = \omega R_1$, $R_3 = \omega^2 R_1$. Verify that R_1 is a cube root of $8(\cos 45° + i \sin 45°)$ by cubing R_1 and applying De Moivre's theorem. Why are the new expressions for R_2 and R_3 evidently also cube roots?

2. Find the three cube roots of -27; those of $-i$; those of ω.

3. Find the two square roots of i; those of $-i$; those of ω.

4. Prove that the numbers $\cos \theta + i \sin \theta$ and no others are represented by points on the circle of radius unity whose center is the origin.

---

[2]Here, as elsewhere when the contrary is not specified, zero and negative as well as positive whole numbers are included under the term "integer."

5. If $a+bi$ and $c+di$ are represented by the points A and C in Fig. 3, prove that their sum is represented by the fourth vertex S of the parallelogram two of whose sides are OA and OC. Hence show that the modulus of the sum of two complex numbers is equal to or less than the sum of their moduli, and is equal to or greater than the difference of their moduli.

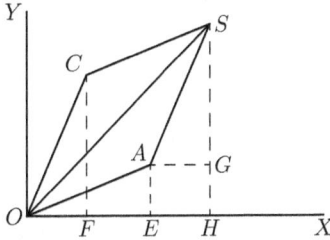   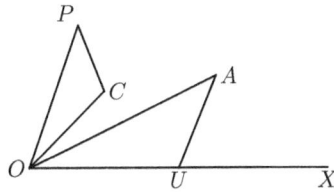

FIG. 3   FIG. 4

6. Let r and r' be the moduli and θ and α the amplitudes of two complex numbers represented by the points A and C in Fig. 4. Let U be the point on the x-axis one unit to the right of the origin O. Construct triangle OCP similar to triangle OUA and similarly placed, so that corresponding sides are OC and OU, CP and UA, OP and OA, while the vertices O, C, P are in the same order (clockwise or counter-clockwise) as the corresponding vertices O, U, A. Prove that P represents the product (§5) of the complex numbers represented by A and C.

7. If $a + bi$ and $e + fi$ are represented by the points A and S in Fig. 3, prove that the complex number obtained by subtracting $a + bi$ from $e + fi$ is represented by the point C. Hence show that the absolute value of the difference of two complex numbers is equal to or less than the sum of their absolute values, and is equal to or greater than the difference of their absolute values.

8. By modifying Ex. 6, show how to construct geometrically the quotient of two complex numbers.

9. nth Roots. As illustrated in §8, it is evident that the nth roots of any complex number $\rho(\cos A + i \sin A)$ are the products of the nth roots of $\cos A + i \sin A$ by the positive real nth root of the positive real number ρ (which may be found by logarithms).

Let an nth root of $\cos A + i \sin A$ be of the form

(4) $r(\cos\theta + i\sin\theta)$.

Then, by De Moivre's theorem,

$$r^n(\cos n\theta + i \sin n\theta) = \cos A + i \sin A.$$

The moduli r^n and 1 must be equal, so that the positive real number r is equal to 1. Since $n\theta$ and A have equal sines and equal cosines, they differ by an integral multiple of 360°. Hence $n\theta = A + k \cdot 360°$, where k is an integer. Substituting the resulting value of θ and the value 1 of r in (4), we get

$$(5) \qquad \cos\left(\frac{A + k \cdot 360°}{n}\right) + i\sin\left(\frac{A + k \cdot 360°}{n}\right).$$

For each integral value of k, (5) is an answer since its nth power reduces to $\cos A + i\sin A$ by DeMoivre's theorem. Next, the value n of k gives the same answer as the value 0 of k; the value $n + 1$ of k gives the same answer as the value 1 of k; and in general the value $n + m$ of k gives the same answer as the value m of k. Hence we may restrict attention to the values $0, 1, \ldots, n - 1$ of k. Finally, the answers (5) given by these values $0, 1, \ldots, n - 1$ of k are all distinct, since they are represented by points whose distance from the origin is the modulus 1 and whose amplitudes are

$$\frac{A}{n}, \qquad \frac{A}{n} + \frac{360°}{n}, \qquad \frac{A}{n} + \frac{2 \cdot 360°}{n}, \ldots, \frac{A}{n} + \frac{(n-1)360°}{n},$$

so that these n points are equally spaced points on a circle of radius unity. Special cases are noted at the end of §10. Hence *any complex number different from zero has exactly n distinct complex nth roots.*

10. Roots of Unity. The trigonometric form of 1 is $\cos 0° + i\sin 0°$. Hence by §9 with $A = 0$, the n distinct nth roots of unity are

$$(6) \qquad \cos\frac{2k\pi}{n} + i\sin\frac{2k\pi}{n} \qquad (k = 0, 1, \ldots, n - 1),$$

where now the angles are measured in radians (an angle of 180 degrees being equal to π radians, where $\pi = 3.1416$, approximately). For $k = 0$, (6) reduces to 1, which is an evident nth root of unity. For $k = 1$, (6) is

$$(7) \qquad R = \cos\frac{2\pi}{n} + i\sin\frac{2\pi}{n}.$$

By De Moivre's theorem, the general number (6) is equal to the kth power of R. Hence the n distinct nth roots of unity are

$$(8) \qquad R, \ R^2, \ R^3, \ldots, \ R^{n-1}, \ R^n = 1.$$

As a special case of the final remark in §9, the n complex numbers (6), and therefore the numbers (8), are represented geometrically by the vertices of a regular polygon of n sides inscribed in the circle of radius unity and center at the origin with one vertex on the positive x-axis.

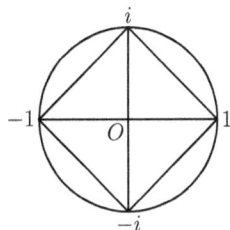

For $n = 3$, the numbers (8) are ω, ω^2, 1, which are represented in Fig. 2 by the vertices of an equilateral triangle.

For $n = 4$, $R = \cos \pi/2 + i \sin \pi/2 = i$. The four fourth roots of unity (8) are i, $i^2 = -1$, $i^3 = -i$, $i^4 = 1$, which are represented by the vertices of a square inscribed in a circle of radius unity and center at the origin O (Fig. 5).

EXERCISES

1. Simplify the trigonometric forms (6) of the four fourth roots of unity. Check the result by factoring $x^4 - 1$.

2. For $n = 6$, show that $R = -\omega^2$. The sixth roots of unity are the three cube roots of unity and their negatives. Check by factoring $x^6 - 1$.

3. From the point representing $a + bi$, how do you obtain that representing $-(a + bi)$? Hence derive from Fig. 2 and Ex. 2 the points representing the six sixth roots of unity. Obtain this result another way.

4. Find the five fifth roots of -1.

5. Obtain the trigonometric forms of the nine ninth roots of unity. Which of them are cube roots of unity?

6. Which powers of a ninth root (7) of unity are cube roots of unity?

11. Primitive nth Roots of Unity.

An nth root of unity is called *primitive* if n is the smallest positive integral exponent of a power of it that is equal to unity. Thus ρ is a primitive nth root of unity if and only if $\rho^n = 1$ and $\rho^l \neq 1$ for all positive integers $l < n$.

Since only the last one of the numbers (8) is equal to unity, the number R, defined by (7), is a primitive nth root of unity. We have shown that the powers (8) of R give all of the nth roots of unity. Which of these powers of R are primitive nth roots of unity?

For $n = 4$, the powers (8) of $R = i$ were seen to be

$$i^1 = i, \quad i^2 = -1, \quad i^3 = -i, \quad i^4 = 1.$$

The first and third are primitive fourth roots of unity, and their exponents 1 and 3 are relatively prime to 4, i.e., each has no divisor > 1 in common with 4. But the second and fourth are not primitive fourth roots of unity (since the square of -1 and the first power of 1 are equal to unity), and their exponents 2 and 4 have the divisor 2 in common with $n = 4$. These facts illustrate and prove the next theorem for the case $n = 4$.

THEOREM.   *The primitive nth roots of unity are those of the numbers* (8) *whose exponents are relatively prime to* n.

Proof.   If k and n have a common divisor d ($d > 1$), R^k is not a primitive nth root of unity, since

$$(R^k)^{\frac{n}{d}} = (R^n)^{\frac{k}{d}} = 1,$$

and the exponent n/d is a positive integer less than n.

But if k and n are relatively prime, i.e., have no common divisor > 1, R^k is a primitive nth root of unity. To prove this, we must show that $(R^k)^l \neq 1$ if l is a positive integer $< n$. By De Moivre's theorem,

$$R^{kl} = \cos \frac{2kl\pi}{n} + i \sin \frac{2kl\pi}{n}.$$

If this were equal to unity, $2kl\pi/n$ would be a multiple of 2π, and hence kl a multiple of n. Since k is relatively prime to n, the second factor l would be a multiple of n, whereas $0 < l < n$.

EXERCISES

1. Show that the primitive cube roots of unity are ω and ω^2.

2. For R given by (7), prove that the primitive nth roots of unity are (i) for $n = 6$, R, R^5; (ii) for $n = 8$, R, R^3, R^5, R^7; (iii) for $n = 12$, R, R^5, R^7, R^{11}.

3. When n is a prime, prove that any nth root of unity, other than 1, is primitive.

4. Let R be a primitive nth root (7) of unity, where n is a product of two different primes p and q. Show that $R, \dots, R^n$ are primitive with the exception of R^p, $R^{2p}, \dots, R^{qp}$, whose qth powers are unity, and R^q, $R^{2q}, \dots, R^{pq}$, whose pth powers are unity. These two sets of exceptions have only R^{pq} in common. Hence there are exactly $pq - p - q + 1$ primitive nth roots of unity.

5. Find the number of primitive nth roots of unity if n is a square of a prime p.

6. Extend Ex. 4 to the case in which n is a product of three distinct primes.

7. If R is a primitive 15th root (7) of unity, verify that R^3, R^6, R^9, R^{12} are the primitive fifth roots of unity, and R^5 and R^{10} are the primitive cube roots of unity. Show that their eight products by pairs give all the primitive 15th roots of unity.

8. If ρ is any primitive nth root of unity, prove that $\rho, \rho^2, \dots, \rho^n$ are distinct and give all the nth roots of unity. Of these show that ρ^k is a primitive nth root of unity if and only if k is relatively prime to n.

9. Show that the six primitive 18th roots of unity are the negatives of the primitive ninth roots of unity.

CHAPTER II

ELEMENTARY THEOREMS ON THE ROOTS OF AN EQUATION

12. Quadratic Equation. If a, b, c are given numbers, $a \neq 0$,

(1) $$ax^2 + bx + c = 0 \quad (a \neq 0)$$

is called a *quadratic equation* or equation of the second degree. The reader is familiar with the following method of solution by "completing the square." Multiply the terms of the equation by $4a$, and transpose the constant term; then

$$4a^2x^2 + 4abx = -4ac.$$

Adding b^2 to complete the square, we get

$$(2ax + b)^2 = \Delta, \qquad \Delta = b^2 - 4ac,$$

(2) $$x_1 = \frac{-b + \sqrt{\Delta}}{2a} \qquad x_2 = \frac{-b - \sqrt{\Delta}}{2a}$$

By addition and multiplication, we find that

(3) $$x_1 + x_2 = \frac{-b}{a}, \qquad x_1 x_2 = \frac{c}{a}.$$

Hence for all values of the variable x,

(4) $$a(x - x_1)(x - x_2) \equiv ax^2 - a(x_1 + x_2)x + ax_1x_2 \equiv ax^2 + bx + c,$$

the sign $\equiv$ being used instead of $=$ since these functions of x are *identically equal*, i.e., the coefficients of like powers of x are the same. We speak of $a(x - x_1)(x - x_2)$ as the *factored form* of the quadratic function $ax^2 + bx + c$, and of $x - x_1$ and $x - x_2$ as its *linear factors*.

In (4) we assign to x the values x_1 and x_2 in turn, and see that

$$0 = ax_1^2 + bx_1 + c, \qquad 0 = ax_2^2 + bx_2 + c.$$

Hence the values (2) are actually the roots of equation (1).

We call $\Delta = b^2 - 4ac$ the *discriminant* of the function $ax^2 + bx + c$ or of the corresponding equation (1). If $\Delta = 0$, the roots (2) are evidently equal, so that, by (4), $ax^2 + bx + c$ is the square of $\sqrt{a}(x - x_1)$, and conversely. We thus obtain the useful result that $ax^2 + bx + c$ *is a perfect square (of a linear function of x) if and only if* $b^2 = 4ac$ (i.e., *if its discriminant is zero*).

Consider a *real* quadratic equation, i.e., one whose coefficients a, b, c are all real numbers. Then if Δ is positive, the two roots (2) are real. But if Δ is negative, the roots are conjugate imaginaries (§2).

When the coefficients of a quadratic equation (1) are any complex numbers, Δ has two complex square roots (§9), so that the roots (2) of (1) are complex numbers, which need not be conjugate.

For example, the discriminant of $x^2 - 2x + c$ is $\Delta = 4(1 - c)$. If $c = 1$, then $\Delta = 0$ and $x^2 - 2x + 1 \equiv (x - 1)^2$ is a perfect square, and the roots 1, 1 of $x^2 - 2x + 1 = 0$ are equal. If $c = 0$, $\Delta = 4$ is positive and the roots 0 and 2 of $x^2 - 2x \equiv x(x - 2) = 0$ are real. If $c = 2$, $\Delta = -4$ is negative and the roots $1 \pm \sqrt{-1}$ of $x^2 - 2x + 2 = 0$ are conjugate complex numbers. The roots of $x^2 - x + 1 + i = 0$ are i and $1 - i$, and are not conjugate.

13. Integral Rational Function, Polynomial.   If n is a positive integer and $c_0, c_1, \ldots, c_n$ are constants (real or imaginary),

$$f(x) \equiv c_0 x^n + c_1 x^{n-1} + \cdots + c_{n-1} x + c_n$$

is called a *polynomial* in x of *degree n*, or also an *integral rational function* of x of degree n. It is given the abbreviated notation $f(x)$, just as the logarithm of $x + 2$ is written $\log(x + 2)$.

If $c_0 \neq 0$, $f(x) = 0$ is an equation of degree n. If $n = 3$, it is often called a *cubic equation*; and, if $n = 4$, a *quartic equation*. For brevity, we often speak of an equation all of whose coefficients are real as a *real equation*.

14. The Remainder Theorem.   *If a polynomial $f(x)$ be divided by $x - c$ until a remainder independent of x is obtained, this remainder is equal to $f(c)$, which is the value of $f(x)$ when $x = c$.*

Denote the remainder by r and the quotient by $q(x)$. Since the dividend is $f(x)$ and the divisor is $x - c$, we have

$$f(x) \equiv (x - c)q(x) + r,$$

identically in x. Taking $x = c$, we obtain $f(c) = r$.

If $r = 0$, the division is exact. Hence we have proved also the following useful theorem.

THE FACTOR THEOREM. *If $f(c)$ is zero, the polynomial $f(x)$ has the factor $x - c$. In other words, if c is a root of $f(x) = 0$, $x - c$ is a factor of $f(x)$.*

For example, 2 is a root of $x^3 - 8 = 0$, so that $x - 2$ is a factor of $x^3 - 8$. Another illustration is furnished by formula (4).

EXERCISES

Without actual division find the remainder when

1. $x^4 - 3x^2 - x - 6$ is divided by $x + 3$.

2. $x^3 - 3x^2 + 6x - 5$ is divided by $x - 3$.

Without actual division show that

3. $18x^{10} + 19x^5 + 1$ is divisible by $x + 1$.

4. $2x^4 - x^3 - 6x^2 + 4x - 8$ is divisible by $x - 2$ and $x + 2$.

5. $x^4 - 3x^3 + 3x^2 - 3x + 2$ is divisible by $x - 1$ and $x - 2$.

6. $r^3 - 1$, $r^4 - 1$, $r^5 - 1$ are divisible by $r - 1$.

7. By performing the indicated multiplication, verify that

$$r^n - 1 \equiv (r - 1)(r^{n-1} + r^{n-2} + \cdots + r + 1).$$

8. In the last identity replace r by x/y, multiply by y^n, and derive

$$x^n - y^n \equiv (x - y)(x^{n-1} + x^{n-2}y + \cdots + xy^{n-2} + y^{n-1}).$$

9. In the identity of Exercise 8 replace y by $-y$, and derive

$$x^n + y^n \equiv (x + y)(x^{n-1} - x^{n-2}y + \cdots - xy^{n-2} + y^{n-1}), \quad n \text{ odd};$$
$$x^n - y^n \equiv (x + y)(x^{n-1} - x^{n-2}y + \cdots + xy^{n-2} - y^{n-1}), \quad n \text{ even}.$$

Verify by the Factor Theorem that $x + y$ is a factor.

10. If a, ar, $ar^2, \ldots, ar^{n-1}$ are n numbers in *geometrical progression* (the ratio of any term to the preceding being a constant $r \neq 1$), prove by Exercise 7 that their sum is equal to

$$\frac{a(r^n - 1)}{r - 1}.$$

11. At the end of each of n years a man deposits in a savings bank a dollars. With annual compound interest at 4%, show that his account at the end of n years will be

$$\frac{a}{.04}\{(1.04)^n - 1\}$$

dollars. Hint: The final deposit draws no interest; the prior deposit will amount to $a(1.04)$ dollars; the deposit preceding that will amount to $a(1.04)^2$ dollars, etc. Hence apply Exercise 10 for $r = 1.04$.

15. Synthetic Division.   The labor of computing the value of a polynomial in x for an assigned value of x may be shortened by a simple device. To find the value of

$$x^4 + 3x^3 - 2x - 5$$

for $x = 2$, note that $x^4 = x \cdot x^3 = 2x^3$, so that the sum of the first two terms of the polynomial is $5x^3$. To $5x^3 = 5 \cdot 2^2 x$ we add the next term $-2x$ and obtain $18x$ or 36. Combining 36 with the final term -5, we obtain the desired value 31.

This computation may be arranged systematically as follows. After supplying zero coefficients of missing powers of x, we write the coefficients in a line, ignoring the powers of x.

1	3	0	-2	-5	$\underline{\quad 2}$
	2	10	20	36	
1	5	10	18	31	

First we bring down the first coefficient 1. Then we multiply it by the given value 2 and enter the product 2 directly under the second coefficient 3, add and write the sum 5 below. Similarly, we enter the product of 5 by 2 under the third coefficient 0, add and write the sum 10 below; etc. The final number 31 in the third line is the value of the polynomial when $x = 2$. The remaining numbers in this third line are the coefficients, in their proper order, of the quotient

$$x^3 + 5x^2 + 10x + 18,$$

which would be obtained by the ordinary long division of the given polynomial by $x - 2$.

We shall now prove that this process, called *synthetic division*, enables us to find the quotient and remainder when any polynomial $f(x)$ is divided by $x - c$. Write

$$f(x) \equiv a_0 x^n + a_1 x^{n-1} + \cdots + a_n,$$

and let the constant remainder be r and the quotient be

$$q(x) \equiv b_0 x^{n-1} + b_1 x^{n-2} + \cdots + b_{n-1}.$$

By comparing the coefficients of $f(x)$ with those in

$$(x - c)q(x) + r \equiv b_0 x^n + (b_1 - cb_0)x^{n-1}$$
$$+ (b_2 - cb_1)x^{n-2} + \cdots + (b_{n-1} - cb_{n-2})x + r - cb_{n-1},$$

we obtain relations which become, after transposition of terms,

$$b_0 = a_0, \quad b_1 = a_1 + cb_0, \quad b_2 = a_2 + cb_1, \ldots, \quad b_{n-1} = a_{n-1} + cb_{n-2}, \quad r = a_n + cb_{n-1}.$$

The steps in the work of computing the b's may be tabulated as follows:

a_0	a_1	a_2	$\cdots$	a_{n-1}	a_n	c
	cb_0	cb_1	$\cdots$	cb_{n-2}	cb_{n-1}	
b_0	b_1	b_2	$\cdots$	$b_{n-1},$	r	

In the second space below a_0 we write b_0 (which is equal to a_0). We multiply b_0 by c and enter the product directly under a_1, add and write the sum b_1 below it. Next we multiply b_1 by c and enter the product directly under a_2, add and write the sum b_2 below it; etc.

EXERCISES

Work each of the following exercises by synthetic division.

1. Divide $x^3 + 3x^2 - 2x - 5$ by $x - 2$.

2. Divide $2x^5 - x^3 + 2x - 1$ by $x + 2$.

3. Divide $x^3 + 6x^2 + 10x - 1$ by $x - 0.09$.

4. Find the quotient of $x^3 - 5x^2 - 2x + 24$ by $x - 4$, and then divide the quotient by $x - 3$. What are the roots of $x^3 - 5x^2 - 2x + 24 = 0$?

5. Given that $x^4 - 2x^3 - 7x^2 + 8x + 12 = 0$ has the roots -1 and 2, find the quadratic equation whose roots are the remaining two roots of the given equation, and find these roots.

6. If $x^4 - 2x^3 - 12x^2 + 10x + 3 = 0$ has the roots 1 and -3, find the remaining two roots.

7. Find the quotient of $2x^4 - x^3 - 6x^2 + 4x - 8$ by $x^2 - 4$.

8. Find the quotient of $x^4 - 3x^3 + 3x^2 - 3x + 2$ by $x^2 - 3x + 2$.

9. Solve Exercises 1, 2, 3, 6, 7 of §14 by synthetic division.

16. Factored Form of a Polynomial.   Consider a polynomial

$$f(x) \equiv c_0 x^n + c_1 x^{n-1} + \cdots + c_n \quad (c_0 \neq 0),$$

whose leading coefficient c_0 is not zero. If $f(x) = 0$ has the root α_1, which may be any complex number, the Factor Theorem shows that $f(x)$ has the factor $x - \alpha_1$, so that

$$f(x) \equiv (x - \alpha_1)Q(x), \quad Q(x) \equiv c_0 x^{n-1} + c_1' x^{n-2} + \cdots + c_{\alpha-1}'.$$

If $Q(x) = 0$ has the root α_2, then

$$Q(x) \equiv (x - \alpha_2)Q_1(x), \quad f(x) \equiv (x - \alpha_1)(x - \alpha_2)Q_1(x).$$

If $Q_1(x) = 0$ has the root α_3, etc., we finally get

$$(5) \qquad\qquad f(x) \equiv c_0(x - \alpha_1)(x - \alpha_2) \cdots (x - \alpha_n).$$

We shall deduce several important conclusions from the preceding discussion. First, suppose that the equation $f(x) = 0$ of degree n is known to have n distinct roots $\alpha_1, \ldots, \alpha_n$. In $f(x) \equiv (x - \alpha_1)Q(x)$ take $x = \alpha_2$; then $0 = (\alpha_2 - \alpha_1)Q(\alpha_2)$, whence $Q(\alpha_2) = 0$ and $Q(x) = 0$ has the root α_2. Similarly, $Q_1(x) = 0$ has the root α_3, etc. Thus all of the assumptions (each introduced by an "if") made in the above discussion have been justified and we have the conclusion (5). Hence *if an equation $f(x) = 0$ of degree n has n distinct roots $\alpha_1, \ldots, \alpha_n$, $f(x)$ can be expressed in the factored form* (5).

It follows readily that the equation can not have a root α different from $\alpha_1, \ldots, \alpha_n$. For, if it did, the left member of (5) is zero when $x = \alpha$ and hence one of the factors of the right member must then be zero, say $\alpha - \alpha_j = 0$, whence the root α is equal to α_j. We have now proved the following important result.

THEOREM.   *An equation of degree n cannot have more than n distinct roots.*

17. Multiple Roots.[1]   Equalities may occur among the α's in (5). Suppose that exactly m_1 of the α's (including α_1) are equal to α_1; that $\alpha_2 \neq \alpha_1$, while exactly m_2 of the α's are equal to α_2; etc. Then (5) becomes

$$(6) \quad f(x) \equiv c_0(x - \alpha_1)^{m_1}(x - \alpha_2)^{m_2} \cdots (x - \alpha_k)^{m_k}, \quad m_1 + m_2 + \cdots + m_k = n,$$

where $\alpha_1, \ldots, \alpha_k$ are distinct. We then call α_1 a *root of multiplicity m_1* of $f(x) = 0$, α_2 a root of multiplicity m_2, etc. In other words, α_1 is a root of

---

[1]Multiple roots are treated by calculus in §58.

multiplicity m_1 of $f(x) = 0$ if $f(x)$ is exactly divisible by $(x - \alpha_1)^{m_1}$, but is not divisible by $(x - \alpha_1)^{m_1+1}$. We call α_1 also an m_1-*fold root*. In the particular cases $m_1 = 1$, 2, and 3, we also speak of α_1 as a *simple root, double root*, and *triple root*, respectively. For example, 4 is a simple root, 3 a double root, -2 a triple root, and 6 a root of multiplicity 4 (or a 4-fold root) of the equation

$$7(x - 4)(x - 3)^2(x + 2)^3(x - 6)^4 = 0$$

of degree 10 which has no further root. This example illustrates the next theorem, which follows from (6) exactly as the theorem in §16 followed from (5).

THEOREM. *An equation of degree n cannot have more than n roots, a root of multiplicity m being counted as m roots.*

18. Identical Polynomials. *If two polynomials in x,*

$$a_0x^n + a_1x^{n-1} + \cdots + a_n, \qquad b_0x^n + b_1x^{n-1} + \cdots + b_n,$$

each of degree n, are equal in value for more than n distinct values of x, they are term by term identical, i.e., $a_0 = b_0$, $a_1 = b_1, \ldots, a_n = b_n$.

For, taking their difference and writing $c_0 = a_0 - b_0, \ldots, c_n = a_n - b_n$, we have

$$c_0x^n + c_1x^{n-1} + \cdots + c_n = 0$$

for more than n distinct values of x. If $c_0 \neq 0$, we would have a contradiction with the theorem in §16. Hence $c_0 = 0$. If $c_1 \neq 0$, we would have a contradiction with the same theorem with n replaced by $n - 1$. Hence $c_1 = 0$, etc. Thus $a_0 = b_0$, $a_1 = b_1$, etc.

EXERCISES

1. Find a cubic equation having the roots 0, 1, 2.

2. Find a quartic equation having the roots ± 1, ± 2.

3. Find a quartic equation having the two double roots 3 and -3.

4. Find a quartic equation having the root 2 and the triple root 1.

5. What is the condition that $ax^2 + bx + c = 0$ shall have a double root?

6. If $a_0x^n + \cdots + a_n = 0$ has more than n distinct roots, each coefficient is zero.

7. Why is there a single answer to each of Exercises 1–4, if the coefficient of the highest power of the unknown be taken equal to unity? State and answer the corresponding general question.

19. The Fundamental Theorem of Algebra. *Every algebraic equation with complex coefficients has a complex (real or imaginary) root.*

This theorem, which is proved in the Appendix, implies that *every equation of degree n has exactly n roots if a root of multiplicity m be counted as m roots.* In other words, *every integral rational function of degree n is a product of n linear factors.* For, in §16, equations $f(x) = 0$, $Q(x) = 0$, $Q_1(x) = 0, \ldots$ each has a root, so that (5) and (6) hold.

20. Relations between the Roots and the Coefficients. In §12 we found the sum and the product of the two roots of any quadratic equation and then deduced the factored form of the equation. We now apply the reverse process to any equation

$$(7) \qquad f(x) \equiv c_0 x^n + c_1 x^{n-1} + \cdots + c_n = 0 \qquad (c_0 \neq 0),$$

whose factored form is

$$(8) \qquad f(x) \equiv c_0(x - \alpha_1)(x - \alpha_2) \cdots (x - \alpha_n).$$

Our next step is to find the expanded form of this product. The following special products may be found by actual multiplication:

$$(x - \alpha_1)(x - \alpha_2) \equiv x^2 - (\alpha_1 + \alpha_2)x + \alpha_1\alpha_2,$$

$$(x - \alpha_1)(x - \alpha_2)(x - \alpha_3) \equiv x^3 - (\alpha_1 + \alpha_2 + \alpha_3)x^2$$
$$+ (\alpha_1\alpha_2 + \alpha_1\alpha_3 + \alpha_2\alpha_3)x - \alpha_1\alpha_2\alpha_3.$$

These identities are the cases $n = 2$ and $n = 3$ of the following general formula:

$$(9) \quad (x - \alpha_1)(x - \alpha_2) \cdots (x - \alpha_n) \equiv x^n - (\alpha_1 + \cdots + \alpha_n)x^{n-1}$$
$$+ (\alpha_1\alpha_2 + \alpha_1\alpha_3 + \alpha_2\alpha_3 + \cdots + \alpha_{n-1}\alpha_n)x^{n-2}$$
$$- (\alpha_1\alpha_2\alpha_3 + \alpha_1\alpha_2\alpha_4 + \cdots + \alpha_{n-2}\alpha_{n-1}\alpha_n)x^{n-3}$$
$$+ \cdots + (-1)^n\alpha_1\alpha_2\cdots\alpha_n,$$

the quantities in parentheses being described in the theorem below. If we multiply each member of (9) by $x - \alpha_{n+1}$, it is not much trouble to verify that the resulting identity can be derived from (9) by changing n into $n+1$, so that (9) is proved true by mathematical induction. Hence the quotient of (7) by c_0 is term by term identical with (9), so that

$$\alpha_1 + \alpha_2 + \cdots + \alpha_n = -c_1/c_0,$$
$$\alpha_1\alpha_2 + \alpha_1\alpha_3 + \alpha_2\alpha_3 + \cdots + \alpha_{n-1}\alpha_n = c_2/c_0,$$
$$(10) \qquad \alpha_1\alpha_2\alpha_3 + \alpha_1\alpha_2\alpha_4 + \cdots + \alpha_{n-2}\alpha_{n-1}\alpha_n = -c_3/c_0,$$
$$\vdots$$
$$\alpha_1\alpha_2\cdots\alpha_{n-1}\alpha_n = (-1)^n c_n/c_0.$$

These results may be expressed in the following words:

THEOREM.   *If $a_1, \ldots, a_n$ are the roots of equation (7), the sum of the roots is equal to $-c_1/c_0$, the sum of the products of the roots taken two at a time is equal to c_2/c_0, the sum of the products of the roots taken three at a time is equal to $-c_3/c_0$, etc.; finally, the product of all the roots is equal to $(-1)^n c_n/c_0$.*

Since we may divide the terms of our equation (7) by c_0, the essential part of our theorem is contained in the following simpler statement:

COROLLARY.   *In an equation in x of degree n, in which the coefficient of x^n is unity, the sum of the n roots is equal to the negative of the coefficient of x^{n-1}, the sum of the products of the roots two at a time is equal to the coefficient of x^{n-2}, etc.; finally the product of all the roots is equal to the constant term or its negative, according as n is even or odd.*

For example, in a cubic equation having the roots 2, 2, 5, and having unity as the coefficient of x^3, the coefficient of x is $2 \cdot 2 + 2 \cdot 5 + 2 \cdot 5 = 24$.

EXERCISES

1. Find a cubic equation having the roots 1, 2, 3.

2. Find a quartic equation having the double roots 2 and -2.

3. Solve $x^4 - 6x^3 + 13x^2 - 12x + 4 = 0$, which has two double roots.

4. Prove that one root of $x^3 + px^2 + qx + r = 0$ is the negative of another root if and only if $r = pq$.

5. Solve $4x^3 - 16x^2 - 9x + 36 = 0$, given that one root is the negative of another.

6. Solve $x^3 - 9x^2 + 23x - 15 = 0$, given that one root is the triple of another.

7. Solve $x^4 - 6x^3 + 12x^2 - 10x + 3 = 0$, which has a triple root.

8. Solve $x^3 - 14x^2 - 84x + 216 = 0$, whose roots are in geometrical progression, i.e., with a common ratio r [say m/r, m, mr].

9. Solve $x^3 - 3x^2 - 13x + 15 = 0$, whose roots are in arithmetical progression, i.e., with a common difference d [say $m - d$, m, $m + d$].

10. Solve $x^4 - 2x^3 - 21x^2 + 22x + 40 = 0$, whose roots are in arithmetical progression. [Denote them by $c - 3b$, $c - b$, $c + b$, $c + 3b$, with the common difference $2b$].

11. Find a quadratic equation whose roots are the squares of the roots of $x^2 - px + q = 0$.

12. Find a quadratic equation whose roots are the cubes of the roots of $x^2 - px + q = 0$. Hint: $\alpha^3 + \beta^3 = (\alpha + \beta)^3 - 3\alpha\beta(\alpha + \beta)$.

13. If α and β are the roots of $x^2 - px + q = 0$, find an equation whose roots are (i) α^2/β; and β^2/α; (ii) $\alpha^3\beta$ and $\alpha\beta^3$; (iii) $\alpha + 1/\beta$ and $\beta + 1/\alpha$.

14. Find a necessary and sufficient condition that the roots, taken in some order, of $x^3 + px^2 + qx + r = 0$ shall be in geometrical progression.

15. Solve $x^3 - 28x + 48 = 0$, given that two roots differ by 2.

21. Imaginary Roots occur in Pairs. The two roots of a real quadratic equation whose discriminant is negative are conjugate imaginaries (§12). This fact illustrates the following useful result.

THEOREM.   *If an algebraic equation with real coefficients has the root $a + bi$, where a and b are real and $b \neq 0$, it has also the root $a - bi$.*

Let the equation be $f(x) = 0$ and divide $f(x)$ by

$$(11) \qquad (x - a)^2 + b^2 \equiv (x - a - bi)(x - a + bi)$$

until we reach a remainder $rx + s$ whose degree in x is less than the degree of the divisor. Since the coefficients of the dividend and divisor are all real, those of the quotient $Q(x)$ and remainder are real. We have

$$f(x) \equiv Q(x)\{(x - a)^2 + b^2\} + rx + s,$$

identically in x. This identity is true in particular when $x = a + bi$, so that

$$0 = r(a + bi) + s = ra + s + rbi.$$

Since all of the letters, other than i, denote real numbers, we have (§2) $ra + s = 0$, $rb = 0$. But $b \neq 0$. Hence $r = 0$, and then $s = 0$. Hence $f(x)$ is exactly divisible by the function (11), so that $f(x) = 0$ has the root $a - bi$.

The theorem may be applied to the real quotient $Q(x)$. We obtain the

COROLLARY.   *If a real algebraic equation has an imaginary root of multiplicity m, the conjugate imaginary of this root is a root of multiplicity m.*

Counting a root of multiplicity m as m roots, we see that a real equation cannot have an odd number of imaginary roots. Hence by §19, *a real equation of odd degree has at least one real root.*

Of the n linear factors of a real integral rational function of degree n (§19), those having imaginary coefficients may be paired as in (11). Hence *every integral rational function with real coefficients can be expressed as a product of real linear and real quadratic factors.*

EXERCISES

1. Solve $x^3 - 3x^2 - 6x - 20 = 0$, one root being $-1 + \sqrt{-3}$.

2. Solve $x^4 - 4x^3 + 5x^2 - 2x - 2 = 0$, one root being $1 - i$.

3. Find a cubic equation with real coefficients two of whose roots are 1 and $3 + 2i$.

4. If a real cubic equation $x^3 - 6x^2 + \cdots = 0$ has the root $1 + \sqrt{-5}$, what are the remaining roots? Find the complete equation.

5. If an equation with *rational* coefficients has a root $a + \sqrt{b}$, where a and b are rational, but $\sqrt{b}$ is irrational, prove that it has the root $a - \sqrt{b}$. [Use the method of §21.]

6. Solve $x^4 - 4x^3 + 4x - 1 = 0$, one root being $2 + \sqrt{3}$.

7. Solve $x^3 - (4 + \sqrt{3})x^2 + (5 + 4\sqrt{3})x - 5\sqrt{3} = 0$, having the root $\sqrt{3}$.

8. Solve the equation in Ex. 7, given that it has the root $2 + i$.

9. Find a cubic equation with rational coefficients having the roots $\frac{1}{2}, \frac{1}{2} + \sqrt{2}$.

10. Given that $x^4 - 2x^3 - 5x^2 - 6x + 2 = 0$ has the root $2 - \sqrt{3}$, find another root and by means of the sum and the product of the four roots deduce, without division, the quadratic equation satisfied by the remaining two roots.

11. Granted that a certain cubic equation has the root 2 and no real root different from 2, does it have two imaginary roots?

12. Granted that a certain quartic equation has the roots $2 \pm 3i$, and no imaginary roots different from them, does it have two real roots?

13. By means of the proof of Ex. 5, may we conclude as at the end of §21 that every integral rational function with rational coefficients can be expressed as a product of linear and quadratic factors with rational coefficients?

22. Upper Limit to the Real Roots. Any number which exceeds all real roots of a real equation is called an *upper limit to the real roots*. We shall prove two theorems which enable us to find readily upper limits to the real roots. For some equations Theorem I gives a better (smaller) upper limit than Theorem II; for other equations, the reverse is true. Evidently any positive number is an upper limit to the real roots of an equation having no negative coefficients.

THEOREM I.   *If, in a real equation*

$$f(x) \equiv a_0 x^n + a_1 x^{n-1} + \cdots + a_n = 0 \qquad (a_0 > 0),$$

the first negative coefficient is preceded by k *coefficients which are positive or zero, and if* G *denotes the greatest of the numerical values of the negative coefficients, then each real root is less than* $1 + \sqrt[k]{G/a_0}$.

For example, in $x^5 + 4x^4 - 7x^2 - 40x + 1 = 0$, $G = 40$ and $k = 3$ since we must supply the coefficient zero to the missing power x^3. Thus the theorem asserts that each root is less than $1 + \sqrt[3]{40}$ and therefore less than 4.42. Hence 4.42 is an upper limit to the roots.

    Proof.   For positive values of x, $f(x)$ will be reduced in value or remain unchanged if we omit the terms $a_1 x^{n-1}, \ldots, a_{k-1} x^{n-k+1}$ (which are positive or zero), and if we change each later coefficient $a_k, \ldots, a_n$ to $-G$. Hence

$$f(x) \geqq a_0 x^n - G(x^{n-k} + x^{n-k-1} + \cdots + x + 1).$$

But, by Ex. 7 of §14,

$$x^{n-k} + \cdots + x + 1 \equiv \frac{x^{n-k+1} - 1}{x - 1},$$

if $x \neq 1$. Furthermore,

$$a_0 x^n - G\left(\frac{x^{n-k+1} - 1}{x - 1}\right) \equiv \frac{x^{n-k+1}\{a_0 x^{k-1}(x-1) - G\} + G}{x - 1}.$$

Hence, if $x > 1$,

$$f(x) > \frac{x^{n-k+1}\{a_0 x^{k-1}(x-1) - G\}}{x - 1},$$

$$f(x) > \frac{x^{n-k+1}\{a_0(x-1)^k - G\}}{x - 1}.$$

Thus, for $x > 1$, $f(x) > 0$ and x is not a root if $a_0(x-1)^k - G \geqq 0$, which is true if $x \geqq 1 + \sqrt[k]{G/a_0}$.

23. Another Upper Limit to the Roots.

    THEOREM II.   *If, in a real algebraic equation in which the coefficient of the highest power of the unknown is positive, the numerical value of each negative coefficient be divided by the sum of all the positive coefficients which precede it, the greatest quotient so obtained increased by unity is an upper limit to the roots.*

For the example in §22, the quotients are $7/(1+4)$ and $40/5$, so that Theorem II asserts that $1 + 8$ or 9 is an upper limit to the roots. Theorem I gave the better upper limit 4.42. But for $x^3 + 8x^2 - 9x + c^2 = 0$, Theorem I gives the upper limit 4, while Theorem II gives the better upper limit 2.

We first give the proof for the case of the equation

$$f(x) \equiv p_4 x^4 - p_3 x^3 + p_2 x^2 - p_1 x + p_0 = 0$$

in which each p_i is positive. In view of the identities

$$x^4 \equiv (x-1)(x^3 + x^2 + x + 1) + 1, \qquad x^2 \equiv (x-1)(x+1) + 1,$$

$f(x)$ is equal to the sum of the terms

$$\begin{aligned} p_4(x-1)x^3 &+ p_4(x-1)x^2 + p_4(x-1)x + p_4(x-1) + p_4, \\ -\, p_3 x^3 & \qquad\qquad\quad +p_2(x-1)x + p_2(x-1) + p_2, \\ & \qquad\qquad\qquad\qquad\; -\, p_1 x \qquad\qquad\quad + p_0. \end{aligned}$$

If $x > 1$, negative terms occur only in the first and third columns, while the sum of the terms in each of these two columns will be ≥ 0 if

$$p_4(x-1) - p_3 \geq 0, \quad (p_4 + p_2)(x-1) - p_1 \geq 0.$$

Hence $f(x) > 0$ and x is not a root if

$$x \geq 1 + \frac{p_3}{p_4}, \quad x \geq 1 + \frac{p_1}{p_4 + p_2}.$$

This proves the theorem for the present equation.

Next, let $f(x)$ be modified by changing its constant term to $-p_0$. We modify the above proof by employing the sum $(p_4 + p_2)x - p_0$ of all the terms in the corresponding last two columns. This sum will be > 0 if $x > p_0/(p_4 + p_2)$, which is true if

$$x \geq 1 + \frac{p_0}{p_4 + p_2}.$$

To extend this method of proof to the general case

$$f(x) \equiv a_n x^n + \cdots + a_0 \qquad (a_n > 0),$$

we have only to employ suitable general notations. Let the negative coefficients be $a_{k_1}, \ldots, a_{k_t}$, where $k_1 > k_2 > \cdots > k_t$. For each positive integer m which is $\leq n$ and distinct from $k_1, \ldots, k_t$, we replace x^m by the equal value

$$d(x^{m-1} + x^{m-2} + \cdots + x + 1) + 1$$

where $d \equiv x - 1$. Let $F(x)$ denote the polynomial in x, with coefficients involving d, which is obtained from $f(x)$ by these replacements. Let $x > 1$, so that

d is positive. Thus the terms $a_{k_i}x^{k_i}$ are the only negative quantities occurring in $F(x)$. If $k_i > 0$, the terms of $F(x)$ which involve explicitly the power x^{k_i} are $a_{k_i}x^{k_i}$ and the $a_m dx^{k_i}$ for the various positive coefficients a_m which precede a_{k_i}. The sum of these terms will be ≥ 0 if $a_{k_i} + d\sum a_m \geq 0$, i.e., if

$$x \geq 1 + \frac{-a_{k_i}}{\sum a_m}.$$

There is an additional case if $k_t = 0$, i.e., if a_0 is negative. Then the terms of $F(x)$ not involving x explicitly are a_0 and the $a_m(d+1)$ for the various positive coefficients a_m. Their sum, $a_0 + x\sum a_m$, will be > 0 if

$$x > \frac{-a_0}{\sum a_m},$$

which is true if

$$x \geq 1 + \frac{-a_0}{\sum a_m}.$$

EXERCISES

Apply the methods of both §22 and §23 to find an upper limit to the roots of

1. $4x^5 - 8x^4 + 22x^3 + 98x^2 - 73x + 5 = 0$.

2. $x^4 - 5x^3 + 7x^2 - 8x + 1 = 0$.

3. $x^7 + 3x^6 - 4x^5 + 5x^4 - 6x^3 - 7x^2 - 8 = 0$.

4. $x^7 + 2x^5 + 4x^4 - 8x^2 - 32 = 0$.

5. A lower limit to the negative roots of $f(x) = 0$ may be found by applying our theorems to $f(-x) = 0$, i.e., to the equation derived from $f(x) = 0$ by replacing x by $-x$. Find a lower limit to the negative roots in Exs. 2, 3, 4.

6. Prove that every real root of a real equation $f(x) = 0$ is less than $1 + g/a_0$ if $a_0 > 0$, where g denotes the greatest of the numerical values of $a_1, \ldots, a_n$. Hint: if $x > 0$,

$$a_0 x^n + a_1 x^{n-1} + \cdots \geq a_0 x^n - g(x^{n-1} + \cdots + x + 1).$$

Proceed as in §22 with $k = 1$.

7. Prove that $1 + g \div |a_0|$ is an upper limit for the moduli of all complex roots of any equation $f(x) = 0$ with complex coefficients, where g is the greatest of the values $|a_1|, \ldots, |a_n|$, and $|a|$ denotes the modulus of a. Hint: use Ex. 5 of §8.

24. Integral Roots. *For an equation all of whose coefficients are integers, any integral root is an exact divisor of the constant term.*

For, if x is an integer such that

(12) $$a_0 x^n + \cdots + a_{n-1}x + a_n = 0,$$

where the a's are all integers, then, by transposing terms, we obtain

$$x(-a_0 x^{n-1} - \cdots - a_{n-1}) = a_n.$$

Thus x is an exact divisor of a_n since the quotient is the integer given by the quantity in parenthesis.

EXAMPLE 1.   Find all the integral roots of

$$x^3 + x^2 - 3x + 9 = 0.$$

Solution.   The exact divisors of the constant term 9 are ± 1, ± 3, ± 9. By trial, no one of ± 1, 3 is a root. Next, we find that -3 is a root by synthetic division (§15):

$$
\begin{array}{rrrr|r}
1 & 1 & -3 & 9 & \underline{-3} \\
  & -3 & 6 & -9 & \\
\hline
1 & -2 & 3 & 0 &
\end{array}
$$

Hence the quotient is $x^2 - 2x + 3$, which is zero for $x = 1 \pm \sqrt{-2}$. Thus -3 is the only integral root.

When the constant term has numerous exact divisors, some device may simplify the application of the theorem.

EXAMPLE 2.[2]   Find all the integral roots of

$$y^3 + 12y^2 - 32y - 256 = 0.$$

Solution.   Since all the terms except y^3 are divisible by 2, an integral root y must be divisible by 2. Since all the terms except y^3 are now divisible by 2^4, we have $y = 4z$, where z is an integer. Removing the factor 2^6 from the equation in z, we obtain

$$z^3 + 3z^2 - 2z - 4 = 0.$$

An integral root must divide the constant term 4. Hence, if there are any integral roots, they occur among the numbers ± 1, ± 2, ± 4. By trial, -1 is found to be a root:

$$
\begin{array}{rrrr|r}
1 & 3 & -2 & -4 & \underline{-1} \\
  & -1 & -2 & 4 & \\
\hline
1 & 2 & -4 & 0 &
\end{array}
$$

---

[2]This problem is needed for the solution (§48) of a certain quartic equation.

Hence the quotient is $z^2 + 2z - 4$, which is zero for $z = -1 \pm \sqrt{5}$. Thus $y = 4z = -4$ is the only integral root of the proposed equation.

EXERCISES

Find all the integral roots of

1. $x^3 + 8x^2 + 13x + 6 = 0$.   **2.** $x^3 - 5x^2 - 2x + 24 = 0$.

3. $x^3 - 10x^2 + 27x - 18 = 0$.   **4.** $x^4 + 4x^3 + 8x + 32 = 0$.

5. The equation in Ex. 4 of §23.

25. Newton's Method for Integral Roots. In §24 we proved that an integral root x of equation (12) having integral coefficients must be an exact divisor of a_n. Similarly, if we transpose all but the last two terms of (12), we see that $a_{n-1}x + a_n$ must be divisible by x^2, and hence $a_{n-1} + a_n/x$ divisible by x. By transposing all but the last three terms of (12), we see that their sum must be divisible by x^3, and hence $a_{n-2} + (a_{n-1} + a_n/x)/x$ divisible by x. We thus obtain a series of conditions of divisibility which an integral root must satisfy. The final sum $a_0 + a_1/x + \cdots$ must not merely be divisible by x, but be actually zero, since it is the quotient of the function (12) by x^n.

In practice, we must test in turn the various divisors x of a_n. If a chosen x is not a root, that fact will be disclosed by one of the conditions mentioned. Newton's method is quicker than synthetic division since it usually detects early and throws out wrong guesses as to a root, whereas in synthetic division the decision comes only at the final step.

For example, the divisor -3 of the constant term of

$$(13) \qquad\qquad f(x) \equiv x^4 - 9x^3 + 24x^2 - 23x + 15 = 0$$

is not a root since $-23 + 15/(-3) = -28$ is not divisible by -3. To show that none of the tests fails for 3, so that 3 is a root, we may arrange the work systematically as follows:

$$(14) \qquad \begin{array}{rrrrr|l} 1 & -9 & 24 & -23 & 15 & 3 \\ -1 & 6 & -6 & 5 & & \text{(divisor)} \\ \hline 0 & -3 & 18 & -18 & & \end{array}$$

First we divide the final coefficient 15 by 3, place the quotient 5 directly under the coefficient -23, and add. Next, we divide this sum -18 by 3, place the quotient -6

directly under the coefficient 24, and add. After two more such steps we obtain the sum zero, so that 3 is a root.

It is instructive to obtain the preceding process by suitably modifying synthetic division. First, we replace x by $1/y$ in (13), multiply each term by y^4, and obtain

$$15y^4 - 23y^3 + 24y^2 - 9y + 1 = 0.$$

We may test this for the root $y = \frac{1}{3}$, which corresponds to the root $x = 3$ of (13), by ordinary synthetic division:

$$
\begin{array}{rrrrr|l}
15 & -23 & 24 & -9 & 1 & \frac{1}{3} \\
   &   5 & -6 & 6 & -1 & \text{(multiplier)} \\
\hline
15 & -18 & 18 & -3 & 0 &
\end{array}
$$

The coefficients in the last two lines (after omitting 15) are the same as those of the last two lines in (14) read in reverse order. This should be the case since we have here multiplied the same numbers by $\frac{1}{3}$ that we divided by 3 in (14). The numbers in the present third line are the coefficients of the quotient (§15). Since we equate the quotient to zero for the applications, we may replace these coefficients by the numbers in the second line which are the products of the former numbers by $\frac{1}{3}$. The numbers in the second line of (14) are the negatives of the coefficients of the quotient of $f(x)$ by $x - 3$.

EXAMPLE.   Find all the integral roots of equation (13).

Solution.   For a negative value of x, each term is positive. Hence all the real roots are positive. By §23, 10 is an upper limit to the roots. By §24, any integral root is an exact divisor of the constant term 15. Hence the integral roots, if any, occur among the numbers 1, 3, 5. Since $f(1) = 8$, 1 is not a root. By (14), 3 is a root. Proceeding similarly with the quotient by $x - 3$, whose coefficients are the negatives of the numbers in the second line of (14), we find that 5 is a root.

EXERCISES

1. Solve Exs. 1–4 of §24 by Newton's method.

2. Prove that, in extending the process (14) to the general equation (12), we may employ the final equations in §15 with $r = 0$ and write

$$
\begin{array}{rrrrrrr|l}
a_0 & a_1 & a_2 & \cdots & a_{n-2} & a_{n-1} & a_n & c \\
-b_0 & -b_1 & -b_2 & \cdots & -b_{n-2} & -b_{n-1} & & \text{(divisor)} \\
\hline
0 & -cb_0 & -cb_1 & \cdots & -cb_{n-3} & -cb_{n-2} & &
\end{array}
$$

Here the quotient, $-b_{n-1}$, of a_n by c is placed directly under a_{n-1}, and added to it to yield the sum $-cb_{n-2}$, etc.

26. Another Method for Integral Roots. An integral divisor d of the constant term is not a root if $d - m$ is not a divisor of $f(m)$, where m is any chosen integer. For, if d is a root of $f(x) = 0$, then

$$f(x) \equiv (x - d)Q(x),$$

where $Q(x)$ is a polynomial having integral coefficients (§15). Hence $f(m) = (m - d)q$, where q is the integer $Q(m)$.

In the example of §25, take $d = 15$, $m = 1$. Since $f(1) = 8$ is not divisible by $15 - 1 = 14$, 15 is not an integral root.

Consider the more difficult example

$$f(x) \equiv x^3 - 20x^2 + 164x - 400 = 0,$$

whose constant term has many divisors. There is evidently no negative root, while 21 is an upper limit to the roots. The positive divisors less than 21 of $400 = 2^4 5^2$ are $d = 1, 2, 4, 8, 16, 5, 10, 20$. First, take $m = 1$ and note that $f(1) = -255 = -3 \cdot 5 \cdot 17$. The corresponding values of $d - 1$ are 0, 1, 3, 7, 15, 4, 9, 19; of these, 7, 4, 9, 19 are not divisors of $f(1)$, so that $d = 8$, 5, 10 and 20 are not roots. Next, take $m = 2$ and note that $f(2) = -144$ is not divisible by $16 - 2 = 14$. Hence 16 is not a root. Incidentally, $d = 1$ and $d = 2$ were excluded since $f(d) \neq 0$. There remains only $d = 4$, which is a root.

In case there are numerous divisors within the limits to the roots, it is usually a waste of time to list all these divisors. For, if a divisor is found to be a root, it is preferable to employ henceforth the quotient, as was done in the example in §25.

EXERCISES

Find all the integral roots of

1. $x^4 - 2x^3 - 21x^2 + 22x + 40 = 0$.

2. $y^3 - 9y^2 - 24y + 216 = 0$.

3. $x^4 - 23x^3 + 187x^2 - 653x + 936 = 0$.

4. $x^5 + 47x^4 + 423x^3 + 140x^2 + 1213x - 420 = 0$.

5. $x^5 - 34x^3 + 29x^2 + 212x - 300 = 0$.

27. Rational Roots. *If an equation with integral coefficients*

(15) $$c_0 x^n + c_1 x^{n-1} + \cdots + c_{n-1} x + c_n = 0$$

has the rational root a/b, where a and b are integers without a common divisor > 1, then a is an exact divisor of c_n, and b is an exact divisor of c_0.

Insert the value a/b of x and multiply all terms of the equation by b^n. We obtain

$$c_0 a^n + c_1 a^{n-1} b + \cdots + c_{n-1} a b^{n-1} + c_n b^n = 0.$$

Since a divides all the terms preceding the last term, it divides that term. But a has no divisor in common with b^n; hence a divides c_n. Similarly, b divides all the terms after the first term and hence divides c_0.

EXAMPLE.   Find all the rational roots of

$$2x^3 - 7x^2 + 10x - 6 = 0.$$

Solution.   By the theorem, the denominator of any rational root x is a divisor of 2. Hence $y = 2x$ is an integer. Multiplying the terms of our equation by 4, we obtain

$$y^3 - 7y^2 + 20y - 24 = 0.$$

There is evidently no negative root. By either of the tests in §§22, 23, an upper limit to the positive roots of our equation in x is $1 + 7/2$, so that $y < 9$. Hence the only possible values of an integral root y are 1, 2, 3, 4, 6, 8. Since 1 and 2 are not roots, we try 3:

1	-7	20	-24	$\underline{3}$
	-1	4	-8	
0	-3	12		

Hence 3 is a root and the remaining roots satisfy the equation $y^2 - 4y + 8 = 0$ and are $2 \pm 2i$. Thus the only rational root of the proposed equation is $x = 3/2$.

If $c_0 = 1$, then $b = \pm 1$ and a/b is an integer. Hence we have the

COROLLARY.   *Any rational root of an equation with integral coefficients, that of the highest power of the unknown being unity, is an integer.*

Given any equation with integral coefficients

$$a_0 y^n + a_1 y^{n-1} + \cdots + a_n = 0,$$

we multiply each term by a_0^{n-1}, write $a_0 y = x$, and obtain an equation (15) with integral coefficients, in which the coefficient c_0 of x^n is now unity. By the Corollary, each rational root x is an integer. Hence we need only find all the integral roots x and divide them by a_0 to obtain all the rational roots y of the proposed equation.

Frequently it is sufficient (and of course simpler) to set $ky = x$, where k is a suitably chosen integer less than a_0.

EXERCISES

Find all of the rational roots of

1. $y^4 - \frac{40}{3}y^3 + \frac{130}{3}y^2 - 40y + 9 = 0.$ **2.** $6y^3 - 11y^2 + 6y - 1 = 0.$

3. $108y^3 - 270y^2 - 42y + 1 = 0.$ [Use $k = 6.$]

4. $32y^3 - 6y - 1 = 0.$ [Use the least $k.$]

5. $96y^3 - 16y^2 - 6y + 1 = 0.$ **6.** $24y^3 - 2y^2 - 5y + 1 = 0.$

7. $y^3 - \frac{1}{2}y^2 - 2y + 1 = 0.$ **8.** $y^3 - \frac{2}{3}y^2 + 3y - 2 = 0.$

9. Solve Exs. 2–6 by replacing y by $1/x$.

Find the equations whose roots are the products of 6 by the roots of

10. $y^2 - 2y - \frac{1}{3} = 0.$ **11.** $y^3 - \frac{1}{2}y^2 - \frac{1}{3}y + \frac{1}{4} = 0.$

CHAPTER III

28. Impossible Constructions. We shall prove that it is not possible, by the methods of Euclidean geometry, to trisect all angles, or to construct a regular polygon of 7 or 9 sides. The proof, which is beyond the scope of elementary geometry, is based on principles of the theory of equations. Moreover, the discussion will show that a regular polygon of 17 sides can be constructed with ruler and compasses, a fact not suspected during the twenty centuries from Euclid to Gauss.

29. Graphical Solution of a Quadratic Equation. If a and b are constructible, and

(1) $\qquad x^2 - ax + b = 0$

has real coefficients and real roots, the roots can be constructed with ruler and compasses as follows. Draw a circle having as a diameter the line BQ joining the points $B = (0, 1)$ and $Q = (a, b)$ in Fig. 6. Then *the abscissas ON and OM of the points of intersection of this circle with the x-axis are the roots of* (1).

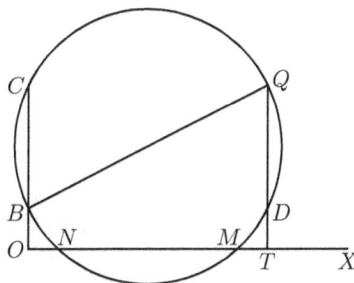

FIG. 6

For, the center of the circle is $(a/2, (b+1)/2)$; the square of BQ is $a^2 + (b-1)^2$; hence the equation of the circle is

$$\left(x - \frac{a}{2}\right)^2 + \left(y - \frac{b+1}{2}\right)^2 = \frac{a^2 + (b-1)^2}{4}.$$

This is found to reduce to (1) when $y = 0$, which proves the theorem.

When the circle is tangent to the x-axis, so that M and N coincide, the two roots are equal. When the circle does not cut the x-axis, or when Q coincides with B, the roots are imaginary.

Another construction follows from §30.

EXERCISES

Solve graphically:

1. $x^2 - 5x + 4 = 0.$ **2.** $x^2 + 5x + 4 = 0.$ **3.** $x^2 + 5x - 4 = 0.$

4. $x^2 - 5x - 4 = 0.$ **5.** $x^2 - 4x + 4 = 0.$ **6.** $x^2 - 3x + 4 = 0.$

30. Analytic Criterion for Constructibility. The first step in our consideration of a problem proposed for construction consists in formulating the problem analytically. In some instances elementary algebra suffices for this formulation. For example, in the ancient problem of the duplication of a cube, we take as a unit of length a side of the given cube, and seek the length x of a side of another cube whose volume is double that of the given cube; hence

$$(2) \qquad\qquad x^3 = 2.$$

But usually it is convenient to employ analytic geometry as in §29; a point is determined by its coordinates x and y with reference to fixed rectangular axes; a straight line is determined by an equation of the first degree, a circle by one of the second degree, in the coordinates of the general point on it. Hence we are concerned with certain numbers, some being the coordinates of points, others being the coefficients of equations, and still others expressing lengths, areas or volumes. These numbers may be said to define analytically the various geometric elements involved.

CRITERION.   *A proposed construction is possible by ruler and compasses if and only if the numbers which define analytically the desired geometric elements can be derived from those defining the given elements by a finite number of rational operations and extractions of real square roots.*

In §29 we were given the numbers a and b, and constructed lines of lengths

$$\tfrac{1}{2}(a \pm \sqrt{a^2 - 4b}).$$

Proof.   First, we grant the condition stated in the criterion and prove that the construction is possible with ruler and compasses. For, a rational function of given quantities is obtained from them by additions, subtractions, multiplications, and divisions. The construction of the sum or difference of two segments is obvious. The construction, by means of parallel lines, of a segment whose length p is equal to the product $a \cdot b$ of the lengths of two given segments is shown in Fig. 7; that for the quotient $q = a/b$ in Fig. 8. Finally, a segment of length $s = \sqrt{n}$ may be constructed, as in Fig. 9, by drawing a semicircle on a diameter composed of two segments of lengths 1 and n, and then drawing a perpendicular to the diameter at the point which separates the two segments. Or we may construct a root of $x^2 - n = 0$ by §29.

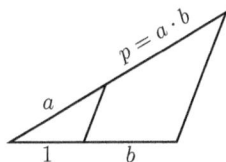

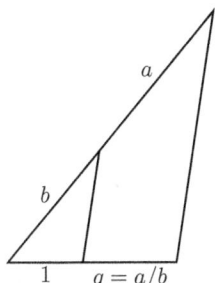

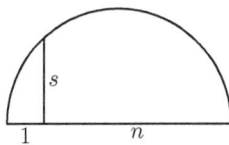

FIG. 7   FIG. 8   FIG. 9

Second, suppose that the proposed construction is possible with ruler and compasses. The straight lines and circles drawn in making the construction are located by means of points either initially given or obtained as the intersections of two straight lines, a straight line and a circle, or two circles. Since the axes of coordinates are at our choice, we may assume that the y-axis is not parallel to any of the straight lines employed in the construction. Then the equation of any one of our lines is

(3) $$y = mx + b.$$

Let $y = m'x + b'$ be the equation of another of our lines which intersects (3). The coordinates of their point of intersection are

$$x = \frac{b' - b}{m - m'}, \qquad y = \frac{mb' - m'b}{m - m'},$$

which are rational functions of the coefficients of the equations of the two lines. Suppose that a line (3) intersects the circle

$$(x - c)^2 + (y - d)^2 = r^2,$$

with the center (c, d) and radius r. To find the coordinates of the points of intersection, we eliminate y between the equations and obtain a quadratic equation for x. Thus x (and hence also $mx + b$ or y) involves no irrationality other than a real square root, besides real irrationalities present in m, b, c, d, r.

Finally, the intersections of two circles are given by the intersections of one of them with their common chord, so that this case reduces to the preceding.

For example, a side of a regular pentagon inscribed in a circle of radius unity is (Ex. 2 of §37)

$$(4) \qquad\qquad s = \tfrac{1}{2}\sqrt{10 - 2\sqrt{5}},$$

which is a number of the type mentioned in the criterion. Hence a regular pentagon can be constructed by ruler and compasses (see the example above quoted).

31. Cubic Equations with a Constructible Root.   We saw that the problem of the duplication of a cube led to a cubic equation (2). We shall later show that each of the problems, to trisect an angle, and to construct regular polygons of 7 and 9 sides with ruler and compasses, leads to a cubic equation. We shall be in a position to treat all of these problems as soon as we have proved the following general result.

THEOREM.   *It is not possible to construct with ruler and compasses a line whose length is a root or the negative of a root of a cubic equation with rational coefficients having no rational root.*

Suppose that x_1 is a root of

$$(5) \qquad\qquad x^3 + \alpha x^2 + \beta x + \gamma = 0 \qquad (\alpha,\ \beta,\ \gamma \text{ rational})$$

such that a line of length x_1 or $-x_1$ can be constructed with ruler and compasses; we shall prove that one of the roots of (5) is rational. We have only to discuss the case in which x_1 is irrational.

By the criterion in §30, since the given numbers in this problem are α, β, γ, all rational, x_1 can be obtained by a finite number of rational operations and extractions of real square roots, performed upon rational numbers or numbers derived from them by such operations. Thus x_1 involves one or more real square roots, but no further irrationalities.

As in the case of (4), there may be superimposed radicals. Such a two-story radical which is not expressible as a rational function, with rational coefficients, of a finite number of square roots of positive rational numbers is said to be a radical of *order* 2. In general, an n-story radical is said to be of order n if it is not expressible as a rational function, with rational coefficients, of radicals each with fewer than n superimposed radicals, the innermost ones affecting positive rational numbers.

We agree to simplify x_1 by making all possible replacements of certain types that are sufficiently illustrated by the following numerical examples.

If x_1 involves $\sqrt{3}$, $\sqrt{5}$, and $\sqrt{15}$, we agree to replace $\sqrt{15}$ by $\sqrt{3} \cdot \sqrt{5}$. If $x_1 = s - 7t$, where s is given by (4) and

$$t = \tfrac{1}{2}\sqrt{10 + 2\sqrt{5}},$$

so that $st = \sqrt{5}$, we agree to write x_1 in the form $s - 7\sqrt{5}/s$, which involves a single radical of order 2 and no new radical of lower order. Finally, we agree to replace $\sqrt{4 - 2\sqrt{3}}$ by its simpler form $\sqrt{3} - 1$.

After all possible simplifications of these types have been made, the resulting expressions have the following properties (to be cited as our agreements): no one of the radicals of highest order n in x_1 is equal to a rational function, with rational coefficients, of the remaining radicals of order n and the radicals of lower orders, while no one of the radicals of order $n - 1$ is equal to a rational function of the remaining radicals of order $n - 1$ and the radicals of lower orders, etc.

Let $\sqrt{k}$ be a radical of highest order n in x_1. Then

$$x_1 = \frac{a + b\sqrt{k}}{c + d\sqrt{k}},$$

where a, b, c, d do not involve $\sqrt{k}$, but may involve other radicals. If $d = 0$, then $c \neq 0$ and we write e for a/c, f for b/c, and get

$$(6) \qquad\qquad x_1 = e + f\sqrt{k}, \qquad\qquad (f \neq 0)$$

where neither e nor f involves $\sqrt{k}$. If $d \neq 0$, we derive (6) by multiplying the numerator and denominator of the fraction for x_1 by $c - d\sqrt{k}$, which is not zero since $\sqrt{k} = c/d$ would contradict our above agreements.

By hypothesis, (6) is a root of equation (5). After expanding the powers and replacing the square of $\sqrt{k}$ by k, we see that

$$(7) \qquad (e + f\sqrt{k})^3 + \alpha(e + f\sqrt{k})^2 + \beta(e + f\sqrt{k}) + \gamma = A + B\sqrt{k},$$

where A and B are certain polynomials in e, f, k and the rational numbers α, β, γ. Thus $A + B\sqrt{k} = 0$. If $B \neq 0$, $\sqrt{k} = -A/B$ is a rational function, with rational coefficients, of the radicals, other than $\sqrt{k}$, in x_1, contrary to our agreements. Hence $B = 0$ and therefore $A = 0$.

When $e - f\sqrt{k}$ is substituted for x in the cubic function (5), the result is the left member of (7) with $\sqrt{k}$ replaced by $-\sqrt{k}$, and hence the result is $A - B\sqrt{k}$. But $A = B = 0$. This shows that

$$(8) \qquad\qquad x_2 = e - f\sqrt{k}$$

is a new root of our cubic equation. Since the sum of the three roots is equal to $-\alpha$ by §20, the third root is

$$(9) \qquad\qquad x_3 = -\alpha - x_1 - x_2 = -\alpha - 2e.$$

Now α is rational. If also e is rational, x_3 is a rational root and we have reached our goal. We next make the assumption that e is irrational and show that it leads to a contradiction. Since e is a component part of the constructible root (6), its only irrationalities are square roots. Let $\sqrt{s}$ be one of the radicals of highest order in e. By the argument which led to (6), we may write $e = e' + f'\sqrt{s}$, whence, by (9),

$$(9') \qquad\qquad x_3 = g + h\sqrt{s}, \qquad\qquad (h \neq 0)$$

where neither g nor h involves $\sqrt{s}$. Then by the argument which led to (8), $g - h\sqrt{s}$ is a root, different from x_3, of our cubic equation, and hence is equal to x_1 or x_2 since there are only three roots (§16). Thus

$$g - h\sqrt{s} = e \pm f\sqrt{k}.$$

By definition, $\sqrt{s}$ is one of the radicals occurring in e. Also, by (9'), every radical occurring in g or h occurs in x_3 and hence in $e = \frac{1}{2}(-\alpha - x_3)$, by (9), α being rational. Hence $\sqrt{k}$ is expressible rationally in terms of the remaining radicals occurring in e and f, and hence in x_1, whose value is given by (6). But this contradicts one of our agreements.

32. Trisection of an Angle. For a given angle A, we can construct with ruler and compasses a line of length $\cos A$ or $-\cos A$, namely the adjacent leg of a right triangle, with hypotenuse unity, formed by dropping a perpendicular from a point in one side of A to the other, produced if necessary. If it were possible to trisect angle A, i.e., construct the angle $A/3$ with ruler and compasses, we could as before construct a line whose length is $\pm\cos(A/3)$. Hence if we show that this last cannot be done when the only given geometric elements are the angle A and a line of unit length, we shall have proved that the angle A cannot be trisected. We shall give the proof for $A = 120°$.

We employ the trigonometric identity

$$\cos A = 4\cos^3 \frac{A}{3} - 3\cos\frac{A}{3}.$$

Multiply each term by 2 and write x for $2\cos(A/3)$. Thus

$$(10) \qquad\qquad x^3 - 3x = 2\cos A.$$

For $A = 120°$, $\cos A = -\frac{1}{2}$ and (10) becomes

(11) $$x^3 - 3x + 1 = 0.$$

Any rational root is an integer (§27) which is an exact divisor of the constant term (§24). By trial, neither $+1$ nor -1 is a root. Hence (11) has no rational root. Hence (§31) *it is not possible to trisect all angles with ruler and compasses.*

Certain angles, like $90°$, $180°$, can be trisected. When $A = 180°$, the equation (10) becomes $x^3 - 3x = -2$ and has the rational root $x = 1$. It is the rationality of a root which accounts for the possibility of trisecting this special angle $180°$.

33. Regular Polygon of 9 Sides, Duplication of a Cube.

Since angle $120°$ cannot be trisected with ruler and compasses (§32), angle $40°$ cannot be so constructed in terms of angle $120°$ and the line of unit length as the given geometric elements. Since the former of these elements and its cosine are constructible when the latter is given, we may take the line of unit length as the only given element. In a regular polygon of 9 sides, the angle subtended at the center by one side is $\frac{1}{9} \cdot 360° = 40°$. Hence *a regular polygon of 9 sides cannot be constructed with ruler and compasses.* Here, as in similar subsequent statements where the given elements are not specified, the only such element is the line of unit length.

A rational root of $x^3 = 2$ is an integer (§27) which is an exact divisor of 2. The cubes of ± 1 and ± 2 are distinct from 2. Hence there is no rational root. Hence (§§30, 31) *it is not possible to duplicate a cube with ruler and compasses.*

34. Regular Polygon of 7 Sides.

If we could construct with ruler and compasses an angle B containing $360/7$ degrees, we could so construct a line of length $x = 2\cos B$. Since $7B = 360°$, $\cos 3B = \cos 4B$. But

$$2\cos 3B = 2(4\cos^3 B - 3\cos B) = x^3 - 3x,$$
$$2\cos 4B = 2(2\cos^2 2B - 1) = 4(2\cos^2 B - 1)^2 - 2 = (x^2 - 2)^2 - 2.$$

Hence
$$0 = x^4 - 4x^2 + 2 - (x^3 - 3x) = (x - 2)(x^3 + x^2 - 2x - 1).$$

But $x = 2$ would give $\cos B = 1$, whereas B is acute. Hence

(12) $$x^3 + x^2 - 2x - 1 = 0.$$

Since this has no rational root, *it is impossible to construct a regular polygon of 7 sides with ruler and compasses.*

35. Regular Polygon of 7 Sides and Roots of Unity.   If

$$R = \cos\frac{2\pi}{7} + i\sin\frac{2\pi}{7},$$

we saw in §10 that R, R^2, R^3, R^4, R^5, R^6, $R^7 = 1$ give all the roots of $y^7 = 1$ and are complex numbers represented by the vertices of a regular polygon of 7 sides inscribed in a circle of radius unity and center at the origin of coordinates. By §6,

$$\frac{1}{R} = \cos\frac{2\pi}{7} - i\sin\frac{2\pi}{7}, \qquad R + \frac{1}{R} = 2\cos\frac{2\pi}{7}.$$

We saw in §34 that $2\cos(2\pi/7)$ is one of the roots of the cubic equation (12). This equation can be derived in a new manner by utilizing the preceding remarks on 7th roots of unity. Our purpose is not primarily to derive (12) again, but to illustrate some principles necessary in the general theory of the construction of regular polygons.

Removing from $y^7 - 1$ the factor $y - 1$, we get

$$(13) \qquad\qquad y^6 + y^5 + y^4 + y^3 + y^2 + y + 1 = 0,$$

whose roots are R, $R^2, \ldots, R^6$. Since we know that $R + 1/R$ is one of the roots of the cubic equation (12), it is a natural step to make the substitution

$$(14) \qquad\qquad y + \frac{1}{y} = x$$

in (13). After dividing its terms by y^3, we have

$$(13') \qquad\qquad \left(y^3 + \frac{1}{y^3}\right) + \left(y^2 + \frac{1}{y^2}\right) + \left(y + \frac{1}{y}\right) + 1 = 0.$$

By squaring and cubing the members of (14), we see that

$$(15) \qquad\qquad y^2 + \frac{1}{y^2} = x^2 - 2, \qquad y^3 + \frac{1}{y^3} = x^3 - 3x.$$

Substituting these values in (13'), we obtain

$$(12) \qquad\qquad x^3 + x^2 - 2x - 1 = 0.$$

That is, the substitution (14) converts equation (13) into (12).

If in (14) we assign to y the six values $R, \ldots, R^6$, we obtain only three distinct values of x:

$$(16)\ \ x_1 = R + \frac{1}{R} = R + R^6, \quad x_2 = R^2 + \frac{1}{R^2} = R^2 + R^5, \quad x_3 = R^3 + \frac{1}{R^3} = R^3 + R^4.$$

In order to illustrate a general method of the theory of regular polygons, we start with the preceding sums of the six roots in pairs and find the cubic equation having these sums as its roots. For this purpose we need to calculate

$$x_1 + x_2 + x_3, \qquad x_1x_2 + x_1x_3 + x_2x_3, \qquad x_1x_2x_3.$$

First, by (16),

$$x_1 + x_2 + x_3 = R + R^2 + \cdots + R^6 = -1,$$

since $R, \ldots, R^6$ are the roots of (13). Similarly,

$$x_1x_2 + x_1x_3 + x_2x_3 = 2(R + R^2 + \cdots + R^6) = -2,$$
$$x_1x_2x_3 = 2 + R + R^2 + \cdots + R^6 = 1.$$

Consequently (§20), the cubic having x_1, x_2, x_3 as roots is (12).

36. Reciprocal Equations.   Any algebraic equation such that the reciprocal of each root is itself a root of the same multiplicity is called a *reciprocal equation.*

The equation $y^7 - 1 = 0$ is a reciprocal equation, since if r is any root, $1/r$ is evidently also a root. Since (13) has the same roots as this equation, with the exception of unity which is its own reciprocal, (13) is also a reciprocal equation.

If r is any root $\neq 0$ of any equation

$$f(y) \equiv y^n + \cdots + c = 0,$$

$1/r$ is a root of $f(1/y) = 0$ and hence of

$$y^n f\left(\frac{1}{y}\right) \equiv 1 + \cdots + cy^n = 0.$$

If the former is a reciprocal equation, it has also the root $1/r$, so that every root of the former is a root of the latter equation. Hence, by §18, the left member of the latter is identical with $cf(y)$. Equating the constant terms, we have $c^2 = 1$, $c = \pm 1$. Hence

(17) $$y^n f\left(\frac{1}{y}\right) \equiv \pm f(y).$$

Thus if $p_i y^{n-i}$ is a term of $f(y)$, also $\pm p_i y^i$ is a term. Hence

(18') $$f(y) \equiv y^n \pm 1 + p_1(y^{n-1} \pm y) + p_2(y^{n-2} \pm y^2) + \cdots .$$

If n is *odd*, $n = 2t + 1$, the final term is $p_t(y^{t+1} \pm y^t)$, and $y \pm 1$ is a factor of $f(y)$. In view of (17), the quotient

$$Q(y) \equiv \frac{f(y)}{y \pm 1}$$

has the property that

$$y^{n-1}Q\left(\frac{1}{y}\right) \equiv Q(y).$$

Comparing this with (17), which implied (18'), we see that $Q(y) = 0$ is a reciprocal equation of the type

$$(18)\quad y^{2t} + 1 + c_1(y^{2t-1} + y) + c_2(y^{2t-2} + y^2) + \cdots + c_{t-1}(y^{t+1} + y^{t-1}) + c_t y^t = 0.$$

If n is *even*, $n = 2t$, and if the upper sign holds in (17), then (18') is of the form (18). Next, let the lower sign hold in (17). Since a term $p_t y^t$ would imply a term $-p_t y^t$, we have $p_t = 0$. The final term in (18') is therefore $p_{t-1}(y^{t+1} - y^{t-1})$. Hence $f(y)$ has the factor $y^2 - 1$. The quotient $q(y) \equiv f(y)/(y^2 - 1)$ has the property that

$$y^{n-2}q\left(\frac{1}{y}\right) \equiv q(y).$$

Comparing this with (17) as before, we see that $q(y) = 0$ is of the form (18) where now $2t = n - 2$. Hence, at least after removing one or both of the factors $y \pm 1$, *any reciprocal equation may be given the form* (18).

The method by which (13) was reduced to a cubic equation may be used to reduce any equation (18) to an equation in x of half the degree. First, we divide the terms of (18) by y^t and obtain

$$\left(y^t + \frac{1}{y^t}\right) + c_1\left(y^{t-1} + \frac{1}{y^{t-1}}\right) + \cdots + c_{t-1}\left(y + \frac{1}{y}\right) + c_t = 0.$$

Next, we perform the substitution (14) by either of the following methods: We may make use of the relation

$$y^k + \frac{1}{y^k} = x\left(y^{k-1} + \frac{1}{y^{k-1}}\right) - \left(y^{k-2} + \frac{1}{y^{k-2}}\right)$$

to compute the values of $y^k + 1/y^k$ in terms of x, starting with the special cases (14) and (15). For example,

$$y^4 + \frac{1}{y^4} = x\left(y^3 + \frac{1}{y^3}\right) - \left(y^2 + \frac{1}{y^2}\right)$$
$$= x(x^3 - 3x) - (x^2 - 2) = x^4 - 4x^2 + 2.$$

Or we may employ the explicit formula (19) of §107 for the sum $y^k + 1/y^k$ of the kth powers of the roots y and $1/y$ of $y^2 - xy + 1 = 0$.

37. Regular Polygon of 9 Sides and Roots of Unity. If

$$R = \cos\frac{2\pi}{9} + i\sin\frac{2\pi}{9},$$

the powers R, R^2, R^4, R^5, R^7, R^8, are the primitive ninth roots of unity (§11). They are therefore the roots of

(19)
$$\frac{y^9 - 1}{y^3 - 1} = y^6 + y^3 + 1 = 0.$$

Dividing the terms of this reciprocal equation by y^3 and applying the second relation (15), we obtain our former cubic equation (11).

EXERCISES

1. Show by (16) that the roots of (12) are $2\cos 2\pi/7$, $2\cos 4\pi/7$, $2\cos 6\pi/7$.

2. The imaginary fifth roots of unity satisfy $y^4 + y^3 + y^2 + y + 1 = 0$, which by the substitution (14) becomes $x^2 + x - 1 = 0$. It has the root

$$R + \frac{1}{R} = 2\cos\frac{2\pi}{5} = \frac{1}{2}(\sqrt{5} - 1).$$

In a circle of radius unity and center O draw two perpendicular diameters AOA', BOB'. With the middle point M of OA' as center and radius MB draw a circle cutting OA at C (Fig. 10). Show that OC and BC are the sides s_{10} and s_5 of the inscribed regular decagon and pentagon respectively. Hints:

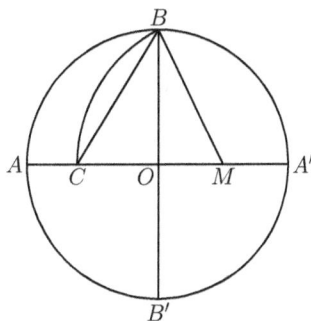

FIG. 10

$$MB = \tfrac{1}{2}\sqrt{5}, \qquad OC = \tfrac{1}{2}(\sqrt{5} - 1), \qquad BC = \sqrt{1 + OC^2} = \tfrac{1}{2}\sqrt{10 - 2\sqrt{5}},$$

$$s_{10} = 2\sin 18° = 2\cos\frac{2\pi}{5} = OC,$$

$$s_5{}^2 = (2\sin 36°)^2 = 2\left(1 - \cos\frac{2\pi}{5}\right) = \frac{1}{4}(10 - 2\sqrt{5}), \qquad s_5 = BC.$$

3. If R is a root of (19) verify as at the end of §35 that $R + R^8$, $R^2 + R^7$, and $R^4 + R^5$ are the roots of (11).

4. Hence show that the roots of (11) are $2\cos 2\pi/9$, $2\cos 4\pi/9$, $2\cos 8\pi/9$.

5. Reduce $y^{11} = 1$ to an equation of degree 5 in x.

6. Solve $y^5 - 7y^4 + y^3 - y^2 + 7y - 1 = 0$ by radicals. [One root is 1.]

7. After finding so easily in Chapter I the trigonometric forms of the complex roots of unity, why do we now go to so much additional trouble to find them algebraically?

8. Prove that every real root of $x^4 + ax^2 + b = 0$ can be constructed with ruler and compasses, given lines of lengths a and b.

9. Show that the real roots of $x^3 - px - q = 0$ are the abscissas of the intersections of the parabola $y = x^2$ and the circle through the origin with the center $(\frac{1}{2}q, \frac{1}{2} + \frac{1}{2}p)$.

Prove that it is impossible, with ruler and compasses:

10. To construct a straight line representing the distance from the circular base of a hemisphere to the parallel plane which bisects the hemisphere.

11. To construct lines representing the lengths of the edges of an existing rectangular parallelopiped having a diagonal of length 5, surface area 24, and volume 1, 2, 3, or 5.

12. To trisect an angle whose cosine is $\frac{1}{2}$, $\frac{1}{3}$, $\frac{1}{4}$, $\frac{1}{8}$ or p/q, where p and q $(q > 1)$ are integers without a common factor, and q is not divisible by a cube.

Prove algebraically that it is possible, with ruler and compasses:

13. To trisect an angle whose cosine is $(4a^3 - 3ab^2)/b^3$, where the integer a is numerically less than the integer b; for example, $\cos^{-1} 11/16$ if $a = -1$, $b = 4$.

14. To construct the legs of a right triangle, given its area and hypotenuse.

15. To construct the third side of a triangle, given two sides and its area.

16. To locate the point P on the side $BC = 1$ of a given square $ABCD$ such that the straight line AP cuts DC produced at a point Q for which the length of PQ is a given number g. Show that $y = BP$ is a root of a reciprocal quartic equation, and solve it when $g = 10$.

38. The Periods of Roots of Unity. Before taking up the regular polygon of 17 sides, we first explain another method of finding the pairs of imaginary seventh roots of unity R and R^6, R^2 and R^5, R^3 and R^4, employed in (16). To this end we seek a positive integer g such that the six roots can be arranged in the order

$$(20) \qquad\qquad R, \quad R^g, \quad R^{g^2}, \quad R^{g^3}, \quad R^{g^4}, \quad R^{g^5},$$

where each term is the gth power of its predecessor. Trying $g = 2$, we find that the fourth term would then be $R^8 = R$. Hence $g \neq 2$. Trying $g = 3$, we obtain

$$(21) \qquad\qquad R, \quad R^3, \quad R^2, \quad R^6, \quad R^4, \quad R^5,$$

where each term is the cube of its predecessor.

To define three *periods*, each of two terms,

$$(16') \qquad\qquad R + R^6, \qquad R^2 + R^5, \qquad R^3 + R^4,$$

we select the first term R of (21) and the third term R^6 after it and add them, then the second term R^3 and the third term R^4 after it, and finally R^2 and the third term R^5 after it.

We may also define two periods, each of three terms,

$$z_1 = R + R^2 + R^4, \qquad z_2 = R^3 + R^6 + R^5,$$

by taking alternate terms in (21).

Since $z_1 + z_2 = -1$, $z_1 z_2 = 3 + R + \cdots + R^6 = 2$, z_1 and z_2 are the roots of $z^2 + z + 2 = 0$. Then R, R^2, R^4 are the roots of $w^3 - z_1 w^2 + z_2 w - 1 = 0$.

39. Regular Polygon of 17 Sides.

Let R be a root $\neq 1$ of $x^{17} = 1$. Then

$$\frac{R^{17} - 1}{R - 1} = R^{16} + R^{15} + \cdots + R + 1 = 0.$$

As in §38, we may take $g = 3$ and arrange the roots $R, \ldots, R^{16}$ so that each is the cube of its predecessor:

$$R, \ R^3, \ R^9, \ R^{10}, \ R^{13}, \ R^5, \ R^{15}, \ R^{11}, \ R^{16}, \ R^{14}, \ R^8, \ R^7, \ R^4, \ R^{12}, \ R^2, \ R^6.$$

Taking alternate terms, we get the two periods, each of eight terms,

$$y_1 = R + R^9 + R^{13} + R^{15} + R^{16} + R^8 + R^4 + R^2,$$
$$y_2 = R^3 + R^{10} + R^5 + R^{11} + R^{14} + R^7 + R^{12} + R^6.$$

Hence $y_1 + y_2 = -1$. We find that $y_1 y_2 = 4(R + \cdots + R^{16}) = -4$. Thus

$$(22) \qquad\qquad y_1, \ y_2 \quad \text{satisfy} \quad y^2 + y - 4 = 0.$$

Taking alternate terms in y_1, we obtain the two periods

$$z_1 = R + R^{13} + R^{16} + R^4, \qquad z_2 = R^9 + R^{15} + R^8 + R^2.$$

Taking alternate terms in y_2, we get the two periods

$$w_1 = R^3 + R^5 + R^{14} + R^{12}, \qquad w_2 = R^{10} + R^{11} + R^7 + R^6.$$

Thus $z_1 + z_2 = y_1$, $w_1 + w_2 = y_2$. We find that $z_1 z_2 = w_1 w_2 = -1$. Hence

(23) z_1, z_2 satisfy $z^2 - y_1 z - 1 = 0$,

(24) w_1, w_2 satisfy $w^2 - y_2 w - 1 = 0$.

Taking alternate terms in z_1, we obtain the periods

$$v_1 = R + R^{16}, \qquad v_2 = R^{13} + R^4.$$

Now, $v_1 + v_2 = z_1$, $v_1 v_2 = w_1$. Hence

(25) v_1, v_2 satisfy $v^2 - z_1 v + w_1 = 0$,

(26) R, R^{16} satisfy $\rho^2 - v_1 \rho + 1 = 0$.

Hence we can find R by solving a series of quadratic equations. Which of the sixteen values of R we shall thus obtain depends upon which root of (22) is called y_1 and which y_2, and similarly in (23)–(26). We shall now show what choice is to be made in each such case in order that we shall finally get the value of the particular root

$$R = \cos \frac{2\pi}{17} + i \sin \frac{2\pi}{17}.$$

Then

$$\frac{1}{R} = \cos \frac{2\pi}{17} - i \sin \frac{2\pi}{17}, \qquad v_1 = R + \frac{1}{R} = 2 \cos \frac{2\pi}{17},$$

$$R^4 = \cos \frac{8\pi}{17} + i \sin \frac{8\pi}{17}, \qquad v_2 = R^4 + \frac{1}{R^4} = 2 \cos \frac{8\pi}{17}.$$

Hence $v_1 > v_2 > 0$, and therefore $z_1 = v_1 + v_2 > 0$. Similarly,

$$w_1 = R^3 + \frac{1}{R^3} + R^5 + \frac{1}{R^5} = 2 \cos \frac{6\pi}{17} + 2 \cos \frac{10\pi}{17} = 2 \cos \frac{6\pi}{17} - 2 \cos \frac{7\pi}{17} > 0,$$

$$y_2 = 2 \cos \frac{6\pi}{17} + 2 \cos \frac{10\pi}{17} + 2 \cos \frac{12\pi}{17} + 2 \cos \frac{14\pi}{17} < 0,$$

since only the first cosine in y_2 is positive and it is numerically less than the third. But $y_1 y_2 = -4$. Hence $y_1 > 0$. Thus (22)–(24) give

$$y_1 = \tfrac{1}{2}(\sqrt{17} - 1), \qquad\qquad y_2 = \tfrac{1}{2}(-\sqrt{17} - 1),$$

$$z_1 = \tfrac{1}{2} y_1 + \sqrt{1 + \tfrac{1}{4} y_1^2}, \qquad\qquad w_1 = \tfrac{1}{2} y_2 + \sqrt{1 + \tfrac{1}{4} y_2^2}.$$

We may readily construct segments of these lengths. Evidently $\sqrt{17}$ is the length of the hypotenuse of a right triangle whose legs are of lengths 1 and 4, while for the radical in z_1 we employ legs of lengths 1 and $\frac{1}{2}y_1$. We thus obtain segments representing the coefficients of the quadratic equation (25). Its roots may be constructed as in §29. The larger root is

$$v_1 = 2\cos\frac{2\pi}{17}.$$

Hence we can construct angle $2\pi/17$ with ruler and compasses, and therefore a regular polygon of 17 sides.

40. Construction of a Regular Polygon of 17 Sides. In a circle of radius unity, construct two perpendicular diameters AB, CD, and draw tangents at A, D, which intersect at S (Fig. 11). Find the point E in AS for which $AE = \frac{1}{4}AS$, by means of two bisections. Then

$$AE = \tfrac{1}{4}, \qquad OE = \tfrac{1}{4}\sqrt{17}.$$

Let the circle with center E and radius OE cut AS at F and F'. Then

$$AF = EF - EA = OE - \tfrac{1}{4} = \tfrac{1}{2}y_1,$$
$$AF' = EF' + EA = OE + \tfrac{1}{4} = -\tfrac{1}{2}y_2,$$
$$OF = \sqrt{OA^2 + AF^2} = \sqrt{1 + \tfrac{1}{4}y_1^2}, \qquad OF' = \sqrt{1 + \tfrac{1}{4}y_2^2}.$$

Let the circle with center F and radius FO cut AS at H, outside of $F'F$; that with center F' and radius $F'O$ cut AS at H' between F' and F. Then

$$AH = AF + FH = AF + OF = \tfrac{1}{2}y_1 + \sqrt{1 + \tfrac{1}{4}y_1^2} = z_1,$$
$$AH' = F'H' - F'A = OF' - AF' = w_1.$$

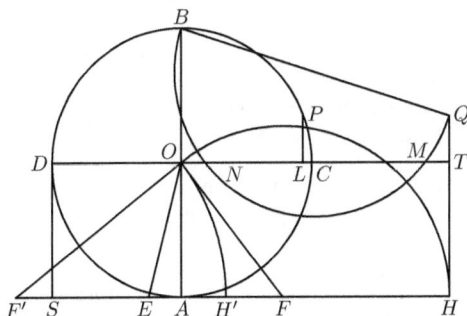

FIG. 11

It remains to construct the roots of equation (25). This will be done as in §29. Draw HTQ parallel to AO and intersecting OC produced at T. Make $TQ = AH'$. Draw a circle having as diameter the line BQ joining $B = (0,1)$ with $Q = (z_1, w_1)$. The abscissas ON and OM of the intersections of this circle with the x-axis OT are the roots of (25). Hence the larger root v_1 is $OM = 2\cos(2\pi/17)$.

Let the perpendicular bisector LP of OM cut the initial circle of unit radius at P. Then

$$\cos LOP = OL = \cos \frac{2\pi}{17}, \qquad LOP = \frac{2\pi}{17}.$$

Hence the chord CP is a side of the inscribed regular polygon of 17 sides, constructed with ruler and compasses.

41. Regular Polygon of n Sides. If n be a prime such that $n-1$ is a power 2^h of 2 (as is the case when $n = 3, 5, 17$), the $n-1$ imaginary nth roots of unity can be separated into 2 sets each of 2^{h-1} roots, each of these sets subdivided into 2 sets each of 2^{h-2} roots, etc., until we reach the pairs R, $1/R$ and R^2, $1/R^2$, etc., and in fact[1] in such a manner that we have a series of quadratic equations, the coefficients of any one of which depend only upon the roots of quadratic equations preceding it in the series. Note that this was the case for $n = 17$ and for $n = 5$. It is in this manner that it can be proved that the roots of $x^n = 1$ can be found in terms of square roots, so that a regular polygon of n sides can be inscribed by ruler and compasses, provided n be a prime of the form $2^h + 1$.

If n be a product of distinct primes of this form, or 2^k times such a product (for example, $n = 15, 30$ or 6), or if $n = 2^m$ $(m > 1)$, it follows readily (see Ex. 1 below) that we can inscribe with ruler and compasses a regular polygon of n sides. But this is impossible for all other values of n.

EXERCISES

1. If a and b are relatively prime numbers, so that their greatest common divisor is unity, we can find integers c and d such that $ac + bd = 1$. Show that, if regular polygons of a and b sides can be constructed and hence angles $2\pi/a$ and $2\pi/b$, a regular polygon of $a \cdot b$ sides can be derived.

2. If $p = 2^h + 1$ is a prime, h is a power of 2. For $h = 2^0, 2^1, 2^2, 2^3$, the values of p are 3, 5, 17, 257 and are primes. [Show that h cannot have an odd factor other than unity.]

---

[1] See the author's article "Constructions with ruler and compasses; regular polygons," in *Monographs on Topics of Modern Mathematics*, Longmans, Green and Co., 1911, p. 374.

3. For 13th roots of unity find the least g (§38), write out the three periods each of four terms, and find the cubic equation having them as roots.

4. For the primitive ninth roots of unity find the least g and write out the three periods each of two terms.

Solve the following reciprocal equations:

5. $y^4 + 4y^3 - 3y^2 + 4y + 1 = 0.$ 　　　　**6.** $y^5 - 4y^4 + y^3 + y^2 - 4y + 1 = 0.$

7. $2y^6 - 5y^5 + 4y^4 - 4y^2 + 5y - 2 = 0.$ 　　　　**8.** $y^5 + 1 = 31(y + 1)^5.$

CHAPTER IV

SOLUTION OF CUBIC AND QUARTIC EQUATIONS; THEIR DISCRIMINANTS

42. Reduced Cubic Equation. If, in the general cubic equation

(1) $$x^3 + bx^2 + cx + d = 0,$$

we set $x = y - b/3$, we obtain the *reduced cubic equation*

(2) $$y^3 + py + q = 0,$$

lacking the square of the unknown y, where

(3) $$p = c - \frac{b^2}{3}, \qquad q = d - \frac{bc}{3} + \frac{2b^3}{27}.$$

After finding the roots y_1, y_2, y_3 of (2), we shall know the roots of (1):

(4) $$x_1 = y_1 - \frac{b}{3}, \qquad x_2 = y_2 - \frac{b}{3}, \qquad x_3 = y_3 - \frac{b}{3}.$$

43. Algebraic Solution of the Reduced Cubic Equation. We shall employ the method which is essentially the same as that given by Vieta in 1591. We make the substitution

(5) $$y = z - \frac{p}{3z}$$

in (2) and obtain

$$z^3 - \frac{p^3}{27z^3} + q = 0,$$

since the terms in z cancel, and likewise the terms in $1/z$. Thus

(6) $$z^6 + qz^3 - \frac{p^3}{27} = 0.$$

Solving this as a quadratic equation for z^3, we obtain

(7) $$z^3 = -\frac{q}{2} \pm \sqrt{R}, \qquad R = \left(\frac{p}{3}\right)^3 + \left(\frac{q}{2}\right)^2.$$

By §8, any number has three cube roots, two of which are the products of the remaining one by the imaginary cube roots of unity:

$$(8) \qquad \omega = -\tfrac{1}{2} + \tfrac{1}{2}\sqrt{3}i, \qquad \omega^2 = -\tfrac{1}{2} - \tfrac{1}{2}\sqrt{3}i.$$

We can choose particular cube roots

$$(9) \qquad A = \sqrt[3]{-\frac{q}{2} + \sqrt{R}}, \qquad B = \sqrt[3]{-\frac{q}{2} - \sqrt{R}},$$

such that $AB = -p/3$, since the product of the numbers under the cube root radicals is equal to $(-p/3)^3$. Hence the six values of z are

$$A, \quad \omega A, \quad \omega^2 A, \quad B, \quad \omega B, \quad \omega^2 B.$$

These can be paired so that the product of the two in each pair is $-p/3$:

$$AB = -\frac{p}{3}, \qquad \omega A \cdot \omega^2 B = -\frac{p}{3}, \qquad \omega^2 A \cdot \omega B = -\frac{p}{3}.$$

Hence with any root z is paired a root equal to $-p/(3z)$. By (5), the sum of the two is a value of y. Hence the *three* values of y are

$$(10) \qquad y_1 = A + B, \qquad y_2 = \omega A + \omega^2 B, \qquad y_3 = \omega^2 A + \omega B.$$

It is easy to verify that these numbers are actually roots of (2). For example, since $\omega^3 = 1$, the cube of y_2 is

$$A^3 + B^3 + 3\omega A^2 B + 3\omega^2 AB^2 = -q - p(\omega A + \omega^2 B) = -q - py_2,$$

by (9) and $AB = -p/3$.

The numbers (10) are known as *Cardan's formulas* for the roots of a reduced cubic equation (2). The expression $A + B$ for a root was first published by Cardan in his *Ars Magna* of 1545, although he had obtained it from Tartaglia under promise of secrecy.

EXAMPLE.   Solve $y^3 - 15y - 126 = 0$.

Solution.   The substitution (5) is here $y = z + 5/z$. We get

$$z^6 - 126z^3 + 125 = 0, \qquad z^3 = 1 \text{ or } 125.$$

The pairs of values of z whose product is 5 are 1 and 5, ω and $5\omega^2$, ω^2 and 5ω. Their sums 6, $\omega + 5\omega^2$, and $\omega^2 + 5\omega$ give the three roots.

EXERCISES

Solve the equations:

1. $y^3 - 18y + 35 = 0.$ **2.** $x^3 + 6x^2 + 3x + 18 = 0.$

3. $y^3 - 2y + 4 = 0.$ **4.** $28x^3 + 9x^2 - 1 = 0.$

44. Discriminant. The product of the squares of the differences of the roots of any equation in which the coefficient of the highest power of the unknown is unity shall be called the *discriminant* of the equation. For the reduced cubic (2), the discriminant is

(11) $(y_1 - y_2)^2(y_1 - y_3)^2(y_2 - y_3)^2 = -4p^3 - 27q^2,$

a result which should be memorized in view of its important applications. It is proved by means of (10) and $\omega^3 = 1$, $\omega^2 + \omega + 1 = 0$, as follows:

$$y_1 - y_2 = (1 - \omega)(A - \omega^2 B), \qquad y_1 - y_3 = (1 - \omega^2)(A - \omega B),$$
$$y_2 - y_3 = (\omega - \omega^2)(A - B),$$
$$(1 - \omega)(1 - \omega^2) = 3, \quad \omega - \omega^2 = \sqrt{3}i.$$

Since $1, \omega, \omega^2$ are the cube roots of unity,

$$(x - 1)(x - \omega)(x - \omega^2) \equiv x^3 - 1,$$

identically in x. Taking $x = A/B$, we see that

$$(A - B)(A - \omega B)(A - \omega^2 B) = A^3 - B^3 = 2\sqrt{R},$$

by (9). Hence

$$(y_1 - y_2)(y_1 - y_3)(y_2 - y_3) = 6\sqrt{3}\sqrt{R}i.$$

Squaring, we get (11), since $-108R = -4p^3 - 27q^2$ by (7). For later use, we note that the discriminant of the reduced cubic is equal to $-108R$.

The discriminant Δ *of the general cubic* (1) *is equal to the discriminant of the corresponding reduced cubic* (2). For, by (4),

$$x_1 - x_2 = y_1 - y_2, \qquad x_1 - x_3 = y_1 - y_3, \qquad x_2 - x_3 = y_2 - y_3.$$

Inserting in (11) the values of p and q given by (3), we get

(12) $\Delta = 18bcd - 4b^3d + b^2c^2 - 4c^3 - 27d^2.$

It is sometimes convenient to employ a cubic equation

(13) $ax^3 + bx^2 + cx + d = 0 \quad (a \neq 0),$

in which the coefficient of x^3 has not been made unity by division. The product P of the squares of the differences of its roots is evidently derived from (12) by replacing b, c, d by $b/a, c/a, d/a$. Hence

(14) $a^4 P = 18abcd - 4b^3d + b^2c^2 - 4ac^3 - 27a^2d^2.$

This expression (and not P itself) is called the discriminant of (13).

45. Number of Real Roots of a Cubic Equation. *A cubic equation with real coefficients has three distinct real roots if its discriminant* Δ *is positive, a single real root and two conjugate imaginary roots if* Δ *is negative, and at least two equal real roots if* Δ *is zero.*

If the roots x_1, x_2, x_3 are all real and distinct, the square of the difference of any two is positive and hence Δ is positive.

If x_1 and x_2 are conjugate imaginaries and hence x_3 is real (§21), $(x_1 - x_2)^2$ is negative. Since $x_1 - x_3$ and $x_2 - x_3$ are conjugate imaginaries, their product is positive. Hence Δ is negative.

If $x_1 = x_2$, Δ is zero. If x_2 were imaginary, its conjugate would be equal to x_3 by §21, and x_2, x_3 would be the roots of a real quadratic equation. The remaining factor $x - x_1$ of the cubic would have real coefficients, whereas $x_1 = x_2$ is imaginary. Hence the equal roots must be real.

Our theorem now follows from these three results by formal logic. For example, if Δ is positive, the roots are all real and distinct, since otherwise either two would be imaginary and Δ would be negative, or two would be equal and Δ would be zero.

EXERCISES

Compute the discriminant and find the number of real roots of

1. $y^3 - 2y - 4 = 0$.   **2.** $y^3 - 15y + 4 = 0$.

3. $y^3 - 27y + 54 = 0$.   **4.** $x^3 + 4x^2 - 11x + 6 = 0$.

5. Show by means of §21 that a double root of a real cubic is real.

46. Irreducible Case. When the roots of a real cubic equation are all real and distinct, the discriminant Δ is positive and $R = -\Delta/108$ is negative, so that Cardan's formulas present the values of the roots in a form involving cube roots of imaginaries. This is called the irreducible case since it may be shown that a cube root of a general complex number cannot be expressed in the form $a + bi$, where a and b involve only real radicals.[1] While we cannot always find these cube roots algebraically, we have learned how to find them trigonometrically (§8).

Example. Solve the cubic equation (2) when $p = -12$, $q = -8\sqrt{2}$.

---

[1] Author's *Elementary Theory of Equations*, pp. 35, 36.

Solution.   By (7), $R = -32$ Hence formulas (9) become

$$A = \sqrt[3]{4\sqrt{2} + 4\sqrt{2}i}, \qquad B = \sqrt[3]{4\sqrt{2} - 4\sqrt{2}i}.$$

The values of A were found in §8. The values of B are evidently the conjugate imaginaries of the values of A. Hence the roots are

$$4\cos 15°, \quad 4\cos 135°, \quad 4\cos 255°.$$

EXERCISES

1. Solve $y^3 - 15y + 4 = 0$.   **2.** Solve $y^3 - 2y - 1 = 0$.

3. Solve $y^3 - 7y + 7 = 0$.   **4.** Solve $x^3 + 3x^2 - 2x - 5 = 0$.

5. Solve $x^3 + x^2 - 2x - 1 = 0$.   **6.** Solve $x^3 + 4x^2 - 7 = 0$.

47. Trigonometric Solution of a Cubic Equation with $\Delta > 0$.   When the roots of a real cubic equation are all real, i.e., if R is negative, they can be computed simultaneously by means of a table of cosines with much less labor than required by Cardan's formulas. To this end we write the trigonometric identity

$$\cos 3A = 4\cos^3 A - 3\cos A$$

in the form

$$z^3 - \tfrac{3}{4}z - \tfrac{1}{4}\cos 3A = 0 \qquad (z = \cos A).$$

In the given cubic $y^3 + py + q = 0$ take $y = nz$; then

$$z^3 + \frac{p}{n^2}z + \frac{q}{n^3} = 0,$$

which will be identical with the former equation in z if

$$n = \sqrt{-\tfrac{4}{3}p}, \quad \cos 3A = -\tfrac{1}{2}q \div \sqrt{-p^3/27}.$$

Since $R = p^3/27 + q^2/4$ is negative, p must be negative, so that n is real and the value of $\cos 3A$ is real and numerically less than unity. Hence we can find $3A$ from a table of cosines. The three values of z are then

$$\cos A, \quad \cos(A + 120°), \quad \cos(A + 240°).$$

Multiplying these by n, we obtain the three roots y correct to a number of decimal places which depends on the tables used.

EXERCISES

1. For $y^3 - 2y - 1 = 0$, show that $n^2 = 8/3$, $\cos 3A = \sqrt{27/32}$, $3A = 23°17'0''$, $\cos A = 0.99084$, $\cos(A + 120°) = -0.61237$, $\cos(A + 240°) = -0.37847$, and that the roots y are 1.61804, -1, -0.61804.

2. Solve Exs. 1, 3, 4, 5, 6 of §46 by trigonometry.

48. Ferrari's Solution of the Quartic Equation. The general quartic equation

$$(15) \qquad x^4 + bx^3 + cx^2 + dx + e = 0,$$

or equation of degree four, becomes after transposition of terms

$$x^4 + bx^3 = -cx^2 - dx - e.$$

The left member contains two of the terms of the square of $x^2 + \frac{1}{2}bx$. Hence by completing the square, we get

$$(x^2 + \tfrac{1}{2}bx)^2 = (\tfrac{1}{4}b^2 - c)x^2 - dx - e.$$

Adding $(x^2 + \frac{1}{2}bx)y + \frac{1}{4}y^2$ to each member, we obtain

$$(16) \qquad (x^2 + \tfrac{1}{2}bx + \tfrac{1}{2}y)^2 = (\tfrac{1}{4}b^2 - c + y)x^2 + (\tfrac{1}{2}by - d)x + \tfrac{1}{4}y^2 - e.$$

The second member is a perfect square of a linear function of x if and only if its discriminant is zero (§12):

$$(\tfrac{1}{2}by - d)^2 - 4(\tfrac{1}{4}b^2 - c + y)(\tfrac{1}{4}y^2 - e) = 0,$$

which may be written in the form

$$(17) \qquad y^3 - cy^2 + (bd - 4e)y - b^2e + 4ce - d^2 = 0.$$

Choose any root y of this *resolvent cubic equation* (17). Then the right member of (16) is the square of a linear function, say $mx + n$. Thus

$$(18) \qquad x^2 + \tfrac{1}{2}bx + \tfrac{1}{2}y = mx + n \quad \text{or} \quad x^2 + \tfrac{1}{2}bx + \tfrac{1}{2}y = -mx - n.$$

The roots of these quadratic equations are the four roots of (16) and hence of the equivalent equation (15). This method of solution is due to Ferrari (1522–1565).

    EXAMPLE.   Solve $x^4 + 2x^3 - 12x^2 - 10x + 3 = 0$.

Solution.   Here $b = 2$, $c = -12$, $d = -10$, $e = 3$. Hence (17) becomes

$$y^3 + 12y^2 - 32y - 256 = 0,$$

which by Ex. 2 of §24 has the root $y = -4$. Our quartic may be written in the form

$$(x^2 + x)^2 = 13x^2 + 10x - 3.$$

Adding $(x^2 + x)(-4) + 4$ to each member, we get

$$(x^2 + x - 2)^2 = 9x^2 + 6x + 1 = (3x + 1)^2,$$
$$x^2 + x - 2 = \pm(3x + 1), \qquad x^2 - 2x - 3 = 0 \text{ or } x^2 + 4x - 1 = 0,$$

whose roots are $3, -1, -2 \pm \sqrt{5}$. As a check, note that the sum of the roots is -2.

EXERCISES

1. Solve $x^4 - 8x^3 + 9x^2 + 8x - 10 = 0$. Note that (17) is $(y - 9)(y^2 - 24) = 0$.

2. Solve $x^4 - 2x^3 - 7x^2 + 8x + 12 = 0$. Since the right member of (16) is $(8 + y)(x^2 - x) + \frac{1}{4}y^2 - 12$, use $y = -8$.

3. Solve $x^4 - 3x^2 + 6x - 2 = 0$.

4. Solve $x^4 - 2x^2 - 8x - 3 = 0$.

5. Solve $x^4 - 10x^2 - 20x - 16 = 0$.

49. Roots of the Resolvent Cubic Equation.   Let y_1 be the root y which was employed in §48. Let x_1 and x_2 be the roots of the first quadratic equation (18), and x_3 and x_4 the roots of the second. Then

$$x_1 x_2 = \tfrac{1}{2}y_1 - n, \qquad x_3 x_4 = \tfrac{1}{2}y_1 + n, \qquad x_1 x_2 + x_3 x_4 = y_1.$$

If, instead of y_1, another root y_2 or y_3 of the resolvent cubic (17) had been employed in §48, quadratic equations different from (18) would have been obtained, such, however, that their four roots are x_1, x_2, x_3, x_4, paired in a new manner. The root which is paired with x_1 is x_2 or x_3 or x_4. It is now plausible that the values of the three y's are

(19) $y_1 = x_1 x_2 + x_3 x_4, \qquad y_2 = x_1 x_3 + x_2 x_4, \qquad y_3 = x_1 x_4 + x_2 x_3.$

To give a more formal proof that the y's given by (19) are the roots of (17), we employ (§20)

$$x_1 + x_2 + x_3 + x_4 = -b, \qquad x_1x_2x_3 + x_1x_2x_4 + x_1x_3x_4 + x_2x_3x_4 = -d,$$
$$x_1x_2 + x_1x_3 + x_1x_4 + x_2x_3 + x_2x_4 + x_3x_4 = c, \qquad x_1x_2x_3x_4 = e.$$

From these four relations we conclude that

$$y_1 + y_2 + y_3 = c,$$
$$y_1y_2 + y_1y_3 + y_2y_3 = (x_1 + x_2 + x_3 + x_4)(x_1x_2x_3 + \cdots + x_2x_3x_4) - 4x_1x_2x_3x_4$$
$$= bd - 4e,$$
$$y_1y_2y_3 = (x_1x_2x_3 + \cdots)^2 + x_1x_2x_3x_4\{(x_1 + \cdots)^2 - 4(x_1x_2 + \cdots)\}$$
$$= d^2 + e(b^2 - 4c).$$

Hence (§20) y_1, y_2, y_3 are the roots of the cubic equation (17).

50. Discriminant.

The discriminant Δ of the quartic equation (15) is defined to be the product of the squares of the differences of its roots:

$$\Delta = (x_1 - x_2)^2(x_1 - x_3)^2(x_1 - x_4)^2(x_2 - x_3)^2(x_2 - x_4)^2(x_3 - x_4)^2.$$

The fact that Δ is equal to the discriminant of the resolvent cubic equation (17) follows at once from (19), by which

$$y_1 - y_2 = (x_1 - x_4)(x_2 - x_3), \qquad y_1 - y_3 = (x_1 - x_3)(x_2 - x_4),$$
$$y_2 - y_3 = (x_1 - x_2)(x_3 - x_4), \qquad (y_1 - y_2)^2(y_1 - y_3)^2(y_2 - y_3)^2 = \Delta.$$

Hence (§44) Δ is equal to the discriminant $-4p^3 - 27q^2$ of the reduced cubic $Y^3 + pY + q = 0$, obtained from (17) by setting $y = Y + c/3$. Thus

(20) $\qquad p = bd - 4e - \frac{1}{3}c^2, \qquad q = -b^2e + \frac{1}{3}bcd + \frac{8}{3}ce - d^2 - \frac{2}{27}c^3.$

THEOREM. *The discriminant of any quartic equation (15) is equal to the discriminant of its resolvent cubic equation and therefore is equal to the discriminant $-4p^3 - 27q^2$ of the corresponding reduced cubic $Y^3 + pY + q = 0$, whose coefficients have the values (20).*

EXERCISES

1. Find the discriminant of $x^4 - 3x^3 + x^2 + 3x - 2 = 0$ and show that the equation has a multiple root.

2. Show by its discriminant that $x^4 - 8x^3 + 22x^2 - 24x + 9 = 0$ has a multiple root.

3. If a real quartic equation has two pairs of conjugate imaginary roots, show that its discriminant Δ is positive. Hence prove that, if $\Delta < 0$, there are exactly two real roots.

4. Hence show that $x^4 - 3x^3 + 3x^2 - 3x + 2 = 0$ has two real and two imaginary roots.

51. Descartes' Solution of the Quartic Equation. Replacing x by $z - b/4$ in the general quartic (15), we obtain the *reduced* quartic equation

$$(21) \qquad\qquad z^4 + qz^2 + rz + s = 0,$$

lacking the term with z^3. We shall prove that we can express the left member of (21) as the product of two quadratic factors

$$(z^2 + 2kz + l)(z^2 - 2kz + m) = z^4 + (l + m - 4k^2)z^2 + 2k(m - l)z + lm.$$

The conditions are

$$l + m - 4k^2 = q, \qquad 2k(m - l) = r, \qquad lm = s.$$

If $k \neq 0$, the first two give

$$2l = q + 4k^2 - \frac{r}{2k}, \qquad 2m = q + 4k^2 + \frac{r}{2k}.$$

Inserting these values in $2l \cdot 2m = 4s$, we obtain

$$(22) \qquad\qquad 64k^6 + 32qk^4 + 4(q^2 - 4s)k^2 - r^2 = 0.$$

The latter may be solved as a cubic equation for k^2. Any root $k^2 \neq 0$ gives a pair of quadratic factors of (21):

$$(23) \qquad\qquad z^2 \pm 2kz + \tfrac{1}{2}q + 2k^2 \mp \frac{r}{4k}.$$

The four roots of these two quadratic functions are the four roots of (21). This method of Descartes (1596–1650) therefore succeeds unless every root of (22) is zero, whence $q = s = r = 0$, so that (12) is the trivial equation $z^4 = 0$.

For example, consider $z^4 - 3z^2 + 6z - 2 = 0$. Then (22) becomes

$$64k^6 - 3 \cdot 32k^4 + 4 \cdot 17k^2 - 36 = 0.$$

The value $k^2 = 1$ gives the factors $z^2 + 2z - 1$, $z^2 - 2z + 2$. Equating these to zero, we find the four roots $-1 \pm \sqrt{2}$, $1 \pm \sqrt{-1}$.

52. Symmetrical Form of Descartes' Solution. To obtain this symmetrical form, we use all three roots k_1^2, k_2^2, k_3^2 of (22). Then

$$k_1^2 + k_2^2 + k_3^2 = -\tfrac{1}{2}q, \qquad k_1^2 k_2^2 k_3^2 = \frac{r^2}{64}.$$

It is at our choice as to which square root of k_1^2 is denoted by $+k_1$ and which by $-k_1$, and likewise as to $\pm k_2$, $\pm k_3$. For our purposes any choice of these signs is suitable provided the choice give

$$(24) \qquad k_1 k_2 k_3 = -\frac{r}{8}.$$

Let $k_1 \neq 0$. The quadratic function (23) is zero for $k = k_1$ if

$$(z \pm k_1)^2 = -\frac{q}{2} - k_1^2 \pm \frac{r}{4k_1} = k_2^2 + k_3^2 \mp \frac{8k_1 k_2 k_3}{4k_1} = (k_2 \mp k_3)^2.$$

Hence the four roots of the quartic equation (21) are

$$(25) \quad k_1 + k_2 + k_3, \qquad k_1 - k_2 - k_3, \qquad -k_1 + k_2 - k_3, \qquad -k_1 - k_2 + k_3.$$

EXERCISES

1. Solve Exs. 4, 5 of §48 by the method of Descartes.

2. By writing y_1, y_2, y_3 for the roots k_1^2, k_2^2, k_3^2 of

$$(26) \qquad 64y^3 + 32qy^2 + 4(q^2 - 4s)y - r^2 = 0,$$

show that the four roots of (21) are the values of

$$(27) \qquad z = \sqrt{y_1} + \sqrt{y_2} + \sqrt{y_3}$$

for all combinations of the square roots for which

$$(28) \qquad \sqrt{y_1} \cdot \sqrt{y_2} \cdot \sqrt{y_3} = -\frac{r}{8}.$$

3. Euler (1707–1783) solved (21) by assuming that it has a root of the form (27). Square (27), transpose the terms free of radicals, square again, replace the last factor of $8\sqrt{y_1 y_2 y_3} \left(\sqrt{y_1} + \sqrt{y_2} + \sqrt{y_3} \right)$ by z, and identify the resulting quartic in z with (21). Show that y_1, y_2, y_3 are the roots of (26) and that relation (28) holds.

4. Find the six differences of the roots (25) and verify that the discriminant Δ of (21) is equal to the quotient of the discriminant of (26) by 4^6.

5. In the theory of the inflexion points of a plane cubic curve there occurs the equation

$$z^4 - Sz^2 - \tfrac{4}{3}Tz - \tfrac{1}{12}S^2 = 0.$$

Show that (26) now becomes

$$\left(y - \frac{S}{6}\right)^3 = C, \qquad C \equiv \left(\frac{T}{6}\right)^2 - \left(\frac{S}{6}\right)^3,$$

and that the roots of the quartic equation are

$$\pm\sqrt{\tfrac{1}{6}S + \sqrt[3]{C}} \pm \sqrt{\tfrac{1}{6}S + \omega\sqrt[3]{C}} \pm \sqrt{\tfrac{1}{6}S + \omega^2\sqrt[3]{C}},$$

where ω is an imaginary cube root of unity and the signs are to be chosen so that the product of the three summands is equal to $+\tfrac{1}{6}T$.

MISCELLANEOUS EXERCISES

1. Find the coordinates of the single real point of intersection of the parabola $y = x^2$ and the hyperbola $xy - 4x + y + 6 = 0$.

2. Show that the abscissas of the points of intersection of $y = x^2$ and $ax^2 - xy + y^2 - x - (a+5)y - 6 = 0$ are the roots of $x^4 - x^3 - 5x^2 - x - 6 = 0$. Compute the discriminant of the latter and show that only two of the four points of intersection are real.

3. Find the coordinates of the two real points in Ex. 2.

4. A right prism of height h has a square base whose side is b and whose diagonal is therefore $b\sqrt{2}$. If v denotes the volume and d a diagonal of the prism, $v = hb^2$ and $d^2 = h^2 + (b\sqrt{2})^2$. Multiply the last equation by h and replace hb^2 by v. Hence $h^3 - d^2h + 2v = 0$. Its discriminant is zero if $d = 3\sqrt{3}$, $v = 27$; find h.

5. Find the admissible values of h in Ex. 4 when $d = 12$, $v = 332.5$.

6. Find a necessary and sufficient condition that quartic equation (15) shall have one root the negative of another root.

Hint: $(x_1 + x_2)(x_3 + x_4) = q - y_1$. Hence substitute q for y in (17).

7. In the study of parabolic orbits occurs the equation

$$\tan \tfrac{1}{2}v + \tfrac{1}{3}\tan^3 \tfrac{1}{2}v = t.$$

Prove that there is a single real root and that it has the same sign as t.

8. In the problem of three astronomical bodies occurs the equation $x^3 + ax + 2 = 0$. Prove that it has three real roots if and only if $a \leqq -3$.

CHAPTER V

THE GRAPH OF AN EQUATION

53. Use of Graphs in the Theory of Equations. To find geometrically the real roots of a real equation $f(x) = 0$, we construct a graph of $y = f(x)$ and measure the distances from the origin O to the intersections of the graph and the x-axis, whose equation is $y = 0$.

For example to find geometrically the real roots of

(1) $$x^2 - 6x - 3 = 0,$$

we equate the left member to y and make a graph of

(1') $$y = x^2 - 6x - 3.$$

We obtain the parabola in Fig. 12. Of the points shown, P has the *abscissa* $x = OQ = 4$ and the *ordinate* $y = -QP = -11$. From the points of intersection of $y = 0$ (the x-axis OX) with the parabola, we obtain the approximate values 6.46 and -0.46 of the roots of (1).

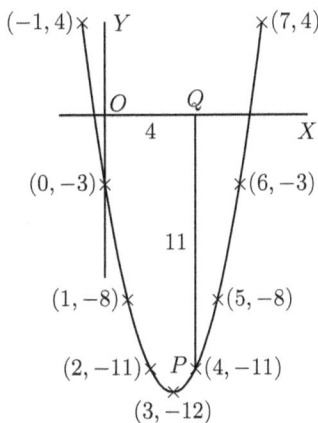

FIG. 12

EXERCISES

1. Find graphically the real roots of $x^2 - 6x + 7 = 0$.

Hint: For each x, $y = x^2 - 6x + 7$ exceeds the y in (1') by 10, so that the new graph is obtained by shifting the parabola in Fig. 12 upward 10 units, leaving the axes OX and OY unchanged. What amounts to the same thing, but is simpler to do, we leave the parabola and OY unchanged, and move the axis OX downward 10 units.

2. Discuss graphically the reality of the roots of $x^2 - 6x + 12 = 0$.

3. Find graphically the roots of $x^2 - 6x + 9 = 0$.

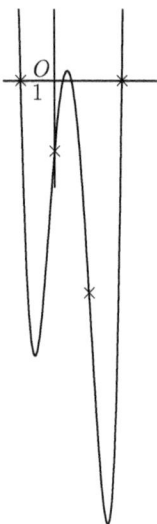

FIG. 13

54. Caution in Plotting. If the example set were

(2) $$y = 8x^4 - 14x^3 - 9x^2 + 11x - 2,$$

one might use successive integral values of x, obtain the points $(-2, 180)$, $(-1, 0)$, $(0, -2)$, $(1, -6)$, $(2, 0)$, $(3, 220)$, all but the first and last of which are shown (by crosses) in Fig. 13, and be tempted to conclude that the graph is a U-shaped curve approximately like that in Fig. 12 and that there are just two real roots, -1 and 2, of

(2') $$8x^4 - 14x^3 - 9x^2 + 11x - 2 = 0.$$

But both of these conclusions would be false. In fact, the graph is a W-shaped curve (Fig. 13) and the additional real roots are $\frac{1}{4}$ and $\frac{1}{2}$.

This example shows that it is often necessary to employ also values of x which are not integers. The purpose of the example was, however, not to point out this obvious fact, but rather to emphasize the chance of serious error in sketching a curve through a number of points, however numerous. The true curve between two points below the x-axis may not cross the x-axis, or may have a peak and actually cross the x-axis twice, or may be an M-shaped curve crossing it four times, etc.

For example, the graph (Fig. 14) of

(3) $$y = x^3 + 4x^2 - 11$$

crosses the x-axis only once; but this fact cannot be established by a graph located by a number of points, however numerous, whose abscissas are chosen at random.

We shall find that correct conclusions regarding the number of real roots may be deduced from a graph whose bend points (§55) have been located.

55. Bend Points. A point (like M or M' in Fig. 14) is called a *bend point* of the graph of $y = f(x)$ if the tangent to the graph at that point is horizontal and if all of the adjacent points of the graph lie below the tangent or all above the tangent. The first, but not the second, condition is satisfied by the point O of

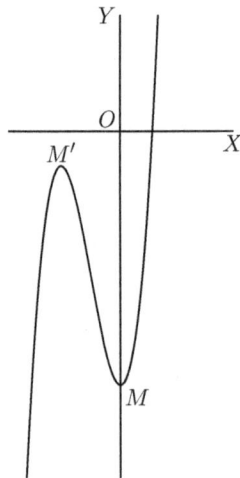

FIG. 14

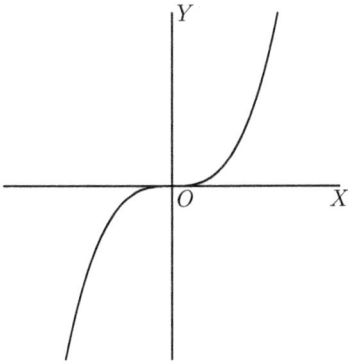

FIG. 15

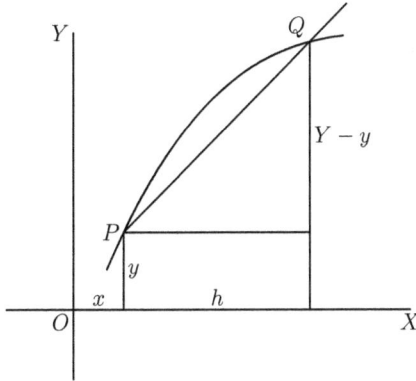

FIG. 16

the graph of $y = x^3$ given in Fig. 15 (see §57). In the language of the calculus, $f(x)$ has a (relative) maximum or minimum value at the abscissa of a bend point on the graph of $y = f(x)$.

Let $P = (x, y)$ and $Q = (x + h, Y)$ be two points on the graph, sketched in Fig. 16, of $y = f(x)$. By the *slope* of a straight line is meant the tangent of the angle between the line and the x-axis, measured counter-clockwise from the latter. In Fig. 16, the slope of the straight line PQ is

(4)
$$\frac{Y - y}{h} = \frac{f(x + h) - f(x)}{h}.$$

For equation (3), $f(x) = x^3 + 4x^2 - 11$. Hence

$$f(x + h) = (x + h)^3 + 4(x + h)^2 - 11$$
$$= x^3 + 4x^2 - 11 + (3x^2 + 8x)h + (3x + 4)h^2 + h^3.$$

The slope (4) of the secant PQ is therefore here

$$3x^2 + 8x + (3x + 4)h + h^2.$$

Now let the point Q move along the graph toward P. Then h approaches the value zero and the secant PQ approaches the tangent at P. The slope of the tangent at P is therefore the corresponding limit $3x^2 + 8x$ of the preceding expression. We call $3x^2 + 8x$ the *derivative* of $x^3 + 4x^2 - 11$.

In particular, if P is a bend point, the slope of the (horizontal) tangent at P is zero, whence $3x^2 + 8x = 0$, $x = 0$ or $x = -\frac{8}{3}$. Equation (3) gives the corresponding values of y. The resulting points

$$M = (0, -11), \qquad M' = (-\tfrac{8}{3}, -\tfrac{41}{27})$$

are easily shown to be bend points. Indeed, for $x > 0$ and for x between -4 and 0, $x^2(x+4)$ is positive, and hence $f(x) > -11$ for such values of x, so that the function (3) has a relative minimum at $x = 0$. Similarly, there is a relative maximum at $x = -\frac{8}{3}$. We may also employ the general method of §59 to show that M and M' are bend points. Since these bend points are both below the x-axis we are now certain that the graph crosses the x-axis only once.

The use of the bend points insures greater accuracy to the graph than the use of dozens of points whose abscissas are taken at random.

56. Derivatives.   We shall now find the slope of the tangent to the graph of $y = f(x)$, where $f(x)$ is any polynomial

(5) $$f(x) = a_0 x^n + a_1 x^{n-1} + \cdots + a_{n-1} x + a_n.$$

We need the expansion of $f(x+h)$ in powers of x. By the binomial theorem,

$$a_0(x+h)^n = a_0 x^n + n a_0 x^{n-1} h + \frac{n(n-1)}{2} a_0 x^{n-2} h^2 + \cdots ,$$

$$a_1(x+h)^{n-1} = a_1 x^{n-1} + (n-1)a_1 x^{n-2} h + \frac{(n-1)(n-2)}{2} a_1 x^{n-3} h^2 + \cdots ,$$

$$\cdots\cdots\cdots\cdots\cdots\cdots\cdots\cdots\cdots\cdots\cdots\cdots\cdots\cdots\cdots\cdots$$

$$a_{n-2}(x+h)^2 = a_{n-2} x^2 + 2a_{n-2} xh + a_{n-2} h^2,$$

$$a_{n-1}(x+h) = a_{n-1} x + a_{n-1} h,$$

$$a_n = a_n.$$

The sum of the left members is evidently $f(x+h)$. On the right, the sum of the first terms (i.e., those free of h) is $f(x)$. The sum of the coefficients of h is denoted by $f'(x)$, the sum of the coefficients of $\frac{1}{2} h^2$ is denoted by $f''(x), \ldots$, the sum of the coefficients of

$$\frac{h^k}{1 \cdot 2 \cdots k}$$

is denoted by $f^{(k)}(x)$. Thus

(6) $$f'(x) = n a_0 x^{n-1} + (n-1)a_1 x^{n-2} + \cdots + 2a_{n-2} x + a_{n-1},$$

(7) $$f''(x) = n(n-1)a_0 x^{n-2} + (n-1)(n-2)a_1 x^{n-3} + \cdots + 2a_{n-2},$$

etc. Hence we have

(8) $$f(x+h) = f(x) + f'(x)h + f''(x)\frac{h^2}{1 \cdot 2} + f'''(x)\frac{h^3}{1 \cdot 2 \cdot 3}$$
$$+ \cdots + f^{(r)}(x)\frac{h^r}{r!} + \cdots + f^{(n)}(x)\frac{h^n}{n!},$$

where $r!$ is the symbol, read r *factorial*, for the product $1 \cdot 2 \cdot 3 \cdots (r-1)r$. Here r is a positive integer, but we include the case $r = 0$ by the definition, $0! = 1$.

This formula (8) is known as *Taylor's theorem* for the present case of a polynomial $f(x)$ of degree n. We call $f'(x)$ the *(first) derivative of $f(x)$*, and $f''(x)$ the *second derivative* of $f(x)$, etc. Concerning the fact that $f''(x)$ is equal to the first derivative of $f'(x)$ and that, in general, the kth derivative $f^{(k)}(x)$ of $f(x)$ is equal to the first derivative of $f^{(k-1)}(x)$, see Exs. 6–9 of the next set.

In view of (8), the limit of (4) as h approaches zero is $f'(x)$. Hence $f'(x)$ *is the slope of the tangent to the graph of $y = f(x)$ at the point (x, y).*

In (5) and (6), let every a be zero except a_0. Thus the derivative of $a_0 x^n$ is $n a_0 x^{n-1}$, and hence is obtained by multiplying the given term by its exponent n and then diminishing its exponent by unity. For example, the derivative of $2x^3$ is $6x^2$.

Moreover, the derivative of $f(x)$ is equal to the sum of the derivatives of its separate terms. Thus the derivative of $x^3 + 4x^2 - 11$ is $3x^2 + 8x$, as found also in §55.

EXERCISES

1. Show that the slope of the tangent to $y = 8x^3 - 22x^2 + 13x - 2$ at (x, y) is $24x^2 - 44x + 13$, and that the bend points are $(0.37, 0.203)$, $(1.46, -5.03)$, approximately. Draw the graph.

2. Prove that the bend points of $y = x^3 - 2x - 5$ are $(.82, -6.09)$, $(-.82, -3.91)$, approximately. Draw the graph and locate the real roots.

3. Find the bend points of $y = x^3 + 6x^2 + 8x + 8$. Locate the real roots.

4. Locate the real roots of $f(x) = x^4 + x^3 - x - 2 = 0$.

Hints: The abscissas of the bend points are the roots of $f'(x) = 4x^3 + 3x^2 - 1 = 0$. The bend points of $y = f'(x)$ are $(0, -1)$ and $(-\frac{1}{2}, -\frac{3}{4})$, so that $f'(x) = 0$ has a single real root (it is just less than $\frac{1}{2}$). The single bend point of $y = f(x)$ is $(\frac{1}{2}, -\frac{37}{16})$, approximately.

5. Locate the real roots of $x^6 - 7x^4 - 3x^2 + 7 = 0$.

6. Prove that $f''(x)$, given by (7), is equal to the first derivative of $f'(x)$.

7. If $f(x) = f_1(x) + f_2(x)$, prove that the kth derivative of f is equal to the sum of the kth derivatives of f_1 and f_2. Use (8).

8. Prove that $f^{(k)}(x)$ is equal to the first derivative of $f^{(k-1)}(x)$. Hint: prove this for $f = ax^m$; then prove that it is true for $f = f_1 + f_2$ if true for f_1 and f_2.

9. Find the third derivative of $x^6 + 5x^4$ by forming successive first derivatives; also that of $2x^5 - 7x^3 + x$.

10. Prove that if g and k are polynomials in x, the derivative of gk is $g'k + gk'$. Hint: multiply the members of $g(x + h) = g(x) + g'(x)h + \cdots$ and $k(x + h) = k(x) + k'(x)h + \cdots$ and use (8) for $f = gk$.

57. Horizontal Tangents. If (x, y) is a bend point of the graph of $y = f(x)$, then, by definition, the slope of the tangent at (x, y) is zero. Hence (§56), the abscissa x is a root of $f'(x) = 0$. In Exs. 1–5 of the preceding set, it was true that, conversely, any real root of $f'(x) = 0$ is the abscissa of a bend point. However, this is not always the case. We shall now consider in detail an example illustrating this fact. The example is the one merely mentioned in §55 to indicate the need of the second requirement made in our definition of a bend point.

The graph (Fig. 15) of $y = x^3$ has no bend point since x^3 increases when x increases. Nevertheless, the derivative $3x^2$ of x^3 is zero for the real value $x = 0$. The tangent to the curve at $(0, 0)$ is the horizontal line $y = 0$. It may be thought of as the limiting position of a secant through O which meets the curve in two further points, seen to be equidistant from O. When one, and hence also the other, of the latter points approaches O, the secant approaches the position of tangency. In this sense the tangent at O is said to meet the curve in three coincident points, their abscissas being the three coinciding roots of $x^3 = 0$. In the language of §17, $x^3 = 0$ has the triple root $x = 0$. The subject of bend points, to which we recur in §59, has thus led us to a digression on the important subject of multiple roots.

58. Multiple Roots. In (8) replace x by α, and h by $x - \alpha$. Then

$$(9) \qquad f(x) = f(\alpha) + f'(\alpha)(x - \alpha) + f''(\alpha)\frac{(x - \alpha)^2}{1 \cdot 2} + f'''(\alpha)\frac{(x - \alpha)^3}{1 \cdot 2 \cdot 3} + \cdots$$
$$+ f^{(m-1)}(\alpha)\frac{(x - \alpha)^{m-1}}{(m - 1)!} + f^{(m)}(\alpha)\frac{(x - \alpha)^m}{m!} + \cdots .$$

By definition (§17) α is a root of $f(x) = 0$ of multiplicity m if $f(x)$ is exactly divisible by $(x - \alpha)^m$, but not by $(x - \alpha)^{m+1}$. Hence α *is a root of multiplicity m of $f(x) = 0$ if and only if*

$$(10) \quad f(\alpha) = 0, \quad f'(\alpha) = 0, \quad f''(\alpha) = 0, \ldots, \quad f^{(m-1)}(\alpha) = 0, \quad f^{(m)}(\alpha) \neq 0.$$

For example, $x^4 + 2x^3 = 0$ has the triple root $x = 0$ since 0 is a root, and since the first and second derivatives $4x^3 + 6x^2$ and $12x^2 + 12x$ are zero for $x = 0$, while the third derivative $24x + 12$ is not zero for $x = 0$.

If in (9) we replace f by f' and hence $f^{(k)}$ by $f^{(k+1)}$, or if we differentiate every term with respect to x, we see by either method that

$$(11) \quad f'(x) = f'(\alpha) + f''(\alpha)(x - \alpha) + \cdots + f^{(m-1)}(\alpha)\frac{(x - \alpha)^{m-2}}{(m - 2)!}$$

$$+ f^{(m)}(\alpha)\frac{(x - \alpha)^{m-1}}{(m - 1)!} + \cdots .$$

Let $f(x)$ and $f'(x)$ have the common factor $(x-\alpha)^{m-1}$, but not the common factor $(x - \alpha)^m$, where $m > 1$. Since (11) has the factor $(x - \alpha)^{m-1}$, we have $f'(\alpha) = 0, \ldots, f^{(m-1)}(\alpha) = 0$. Since also $f(x)$ has the factor $x - \alpha$, evidently $f(\alpha) = 0$. Then, by (9), $f(x)$ has the factor $(x - \alpha)^m$, which, by hypothesis, is not also a factor of $f'(x)$. Hence, in (11), $f^{(m)}(\alpha) \neq 0$. Thus, by (10), α is a root of $f(x) = 0$ of multiplicity m.

Conversely, let α be a root of $f(x) = 0$ of multiplicity m. Then relations (10) hold, and hence, by (11), $f'(x)$ is divisible by $(x - \alpha)^{m-1}$, but not by $(x - \alpha)^m$. Thus $f(x)$ and $f'(x)$ have the common factor $(x - \alpha)^{m-1}$, but not the common factor $(x - \alpha)^m$.

We have now proved the following useful result.

THEOREM. *If $f(x)$ and $f'(x)$ have a greatest common divisor $g(x)$ involving x, a root of $g(x) = 0$ of multiplicity $m - 1$ is a root of $f(x) = 0$ of multiplicity m, and conversely any root of $f(x) = 0$ of multiplicity m is a root of $g(x) = 0$ of multiplicity $m - 1$.*

In view of this theorem, the problem of finding all the multiple roots of $f(x) = 0$ and the multiplicity of each multiple root is reduced to the problem of finding the roots of $g(x) = 0$ and the multiplicity of each.

For example, let $f(x) = x^3 - 2x^2 - 4x + 8$. Then

$$f'(x) = 3x^2 - 4x - 4, \qquad 9f(x) = f'(x)(3x - 2) - 32(x - 2).$$

Since $x - 2$ is a factor of $f'(x)$, it may be taken to be the greatest common divisor of $f(x)$ and $f'(x)$, the choice of the constant factor c in $c(x-2)$ being here immaterial. Hence 2 is a double root of $f(x) = 0$, while the remaining root -2 is a simple root.

EXERCISES

1. Prove that $x^3 - 7x^2 + 15x - 9 = 0$ has a double root.

2. Show that $x^4 - 8x^2 + 16 = 0$ has two double roots.

3. Prove that $x^4 - 6x^2 - 8x - 3 = 0$ has a triple root.

4. Test $x^4 - 8x^3 + 22x^2 - 24x + 9 = 0$ for multiple roots.

5. Test $x^3 - 6x^2 + 11x - 6 = 0$ for multiple roots.

6. Test $x^4 - 9x^3 + 9x^2 + 81x - 162 = 0$ for multiple roots.

59. Ordinary and Inflexion Tangents.   The equation of the straight line through the point (α, β) with the slope s is $y - \beta = s(x - \alpha)$. The slope of the tangent to the graph of $y = f(x)$ at the point (α, β) on it is $s = f'(\alpha)$ by §56. Also, $\beta = f(\alpha)$. Hence the equation of the tangent is

$$(12) \qquad\qquad y = f(\alpha) + f'(\alpha)(x - \alpha).$$

By subtracting the members of this equation from the corresponding members of equation (9), we see that the abscissas x of the points of intersection of the graph of $y = f(x)$ with its tangent satisfy the equation

$$f''(\alpha)\frac{(x - \alpha)^2}{2!} + f'''(\alpha)\frac{(x - \alpha)^3}{3!} + \cdots + f^{(m-1)}(\alpha)\frac{(x - \alpha)^{m-1}}{(m - 1)!}$$
$$+ f^{(m)}(\alpha)\frac{(x - \alpha)^m}{m!} + \cdots = 0.$$

Here the term containing $f^{(m-1)}(\alpha)$ must evidently be suppressed if $m = 2$, since the term containing $f^{(m)}(\alpha)$ then coincides with the first term.

If α is a root of multiplicity m of this equation, i.e., if the left member is divisible by $(x - \alpha)^m$, but not by $(x - \alpha)^{m+1}$, the point (α, β) is counted as m coincident points of intersection of the curve with its tangent (just as in the case of $y = x^3$ and its tangent $y = 0$ in §57). This will be the case if and only if

$$(13) \qquad f''(\alpha) = 0, \qquad f'''(\alpha) = 0, \ldots, \qquad f^{(m-1)}(\alpha) = 0, \qquad f^{(m)}(\alpha) \neq 0,$$

in which $m > 1$ and, as explained above, only the final relation $f''(\alpha) \neq 0$ is retained if $m = 2$. If $m = 3$, the conditions are $f''(\alpha) = 0$, $f^{(3)}(\alpha) \neq 0$.

For example, if $f(x) = x^4$ and $\alpha = 0$, then $f''(0) = f'''(0) = 0$, $f^{(4)}(0) = 24 \neq 0$, so that $m = 4$. The graph of $y = x^4$ is a U-shaped curve, whose intersection with the tangent (the x-axis) at $(0, 0)$ is counted as four coincident points of intersection.

Given $f(x)$ and α, we can find, as in the preceding example, the value of m for which relations (13) hold. We then apply the

THEOREM.   *If m is even $(m > 0)$, the points of the curve in the vicinity of the point of tangency (α, β) are all on the same side of the tangent, which is then called an* **ordinary tangent**. *But if m is odd $(m > 1)$, the curve crosses the tangent at the point of tangency (α, β), and this point is called an* **inflexion point**, *while the tangent is called an* **inflexion tangent**.

For example, in Fig. 15, OX is an inflexion tangent, while the tangent at any point except O is an ordinary tangent. In Figs. 18, 19, 20, the tangents at the points marked by crosses are ordinary tangents, but the tangent at the point midway between them and on the y-axis is an inflexion tangent.

To simplify the proof, we first take as new axes lines parallel to the old axes and intersecting at (α, β). In other words, we set $x - \alpha = X$, $y - \beta = Y$, where X, Y are the coordinates of (x, y) referred to the new axes. Since $\beta = f(\alpha)$, the tangent (12) becomes $Y = f'(\alpha)X$, while, by (9), $y = f(x) = \beta + f'(\alpha)(x-\alpha) + \cdots$ becomes

$$Y = f'(\alpha)X + f''(\alpha)\frac{X^2}{2} + \cdots = f'(\alpha)X + f^{(m)}(\alpha)\frac{X^m}{m!} + \cdots,$$

after omitting terms which are zero by (13).

To simplify further the algebraic work, we pass to oblique axes,[1] the new y-axis coinciding with the Y-axis, while the new x-axis is the tangent, the angle between which and the X-axis is designated by θ. Then

$$\tan \theta = f'(\alpha).$$

By Fig. 17,

$$X = x \cos \theta, \qquad Y - y = f'(\alpha)X.$$

Hence when expressed in terms of the new coordinates x, y, the tangent is $y = 0$, while the equation (14) of the curve becomes

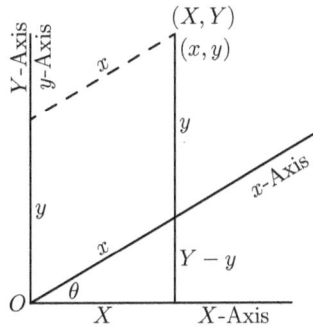

Fig. 17

$$y = cx^m + dx^{m+1} + \cdots, \qquad c = \frac{f^{(m)}(\alpha) \cos^m \theta}{m!} \neq 0.$$

For x sufficiently small numerically, whether positive or negative, the sum of the terms after cx^m is insignificant in comparison with cx^m, so that y has the same sign as cx^m (§64). Hence, if m is even, the points of the curve in the vicinity of the origin and on both sides of it are all on the same side of the x-axis, i.e., the tangent. But, if m is odd, the points with small positive abscissas x lie on one side of the x-axis and those with numerically small negative abscissas lie on the opposite side.

Our transformations of coordinates changed the equations of the curve and of its tangent, but did not change the curve itself and its tangent. Hence our theorem is proved.

---

[1] Since the earlier x, y do not occur in (14) and the new equation of the tangent, we shall designate the final coordinates by x, y without confusion.

By our theorem, α is the abscissa of an inflexion point of the graph of $y = f(x)$ if and only if conditions (13) hold with m odd $(m > 1)$. These conditions include neither $f(\alpha) = 0$ nor $f'(\alpha) = 0$, in contrast with (10). In the theory of equations we are primarily interested in the abscissas α of only those points of inflexion whose inflexion tangents are horizontal, and are interested in them, because we must exclude such roots α of $f'(x) = 0$ when seeking the abscissas of bend points, which are the important points for our purposes. A point on the graph at which the tangent is both horizontal and an ordinary tangent is a bend point by the definition in §55. Hence if we apply our theorem to the special case $f'(\alpha) = 0$, we obtain the following

CRITERION. *Any root α of $f'(x) = 0$ is the abscissa of a bend point of the graph of $y = f(x)$ or of a point with a horizontal inflexion tangent according as the value of m for which relations (13) hold is even or odd.*

For example, if $f(x) = x^4$, then $\alpha = 0$ and $m = 4$, so that $(0,0)$ is a bend point of the U-shaped graph of $y = x^4$. If $f(x) = x^3$, then $\alpha = 0$ and $m = 3$, so that $(0,0)$ is a point with a horizontal inflexion tangent (OX in Fig. 15) of the graph of $y = x^3$.

EXERCISES

1. If $f(x) = 3x^5 + 5x^3 + 4$, the only real root of $f'(x) = 0$ is $x = 0$. Show that $(0,4)$ is an inflexion point, and thus that there is no bend point and hence that $f(x) = 0$ has a single real root.

2. Prove that $x^3 - 3x^2 + 3x + c = 0$ has an inflexion point, but no bend point.

3. Show that $x^5 - 10x^3 - 20x^2 - 15x + c = 0$ has two bend points and no horizontal inflexion tangents.

4. Prove that $3x^5 - 40x^3 + 240x + c = 0$ has no bend point, but has two horizontal inflexion tangents.

5. Prove that any function $x^3 - 3\alpha x^2 + \cdots$ of the third degree can be written in the form $f(x) = (x - \alpha)^3 + ax + b$. The straight line having the equation $y = ax + b$ meets the graph of $y = f(x)$ in three coincident points with the abscissa α and hence is an inflexion tangent. If we take new axes of coordinates parallel to the old and intersecting at the new origin $(\alpha, 0)$, i.e., if we make the transformation $x = X + \alpha$, $y = Y$, of coordinates, we see that the equation $f(x) = 0$ becomes a reduced cubic equation $X^3 + pX + q = 0$ (§42).

6. Find the inflexion tangent to $y = x^3 + 6x^2 - 3x + 1$ and transform $x^3 + 6x^2 - 3x + 1 = 0$ into a reduced cubic equation.

60. Real Roots of a Real Cubic Equation.

It suffices to consider

$$f(x) = x^3 - 3lx + q \qquad (l \neq 0),$$

in view of Ex. 5 above. Then $f' = 3(x^2 - l)$, $f'' = 6x$. If $l < 0$, there is no bend point and the cubic equation $f(x) = 0$ has a single real root. If $l > 0$, there are two bend points

$$(\sqrt{l}, q - 2l\sqrt{l}), \qquad (-\sqrt{l}, q + 2l\sqrt{l}),$$

which are shown by crosses in Figs. 18–20 for the graph of $y = f(x)$ in the three possible cases specified by the inequalities shown below the figures. For a large positive x, the term x^3 in $f(x)$ predominates, so that the graph contains a point high up in the first quadrant, thence extends downward to the right-hand bend point, then ascends to the left-hand bend point, and finally descends. As a check, the graph contains a point far down in the third quadrant, since for x negative, but sufficiently large numerically, the term x^3 predominates and the sign of y is negative.

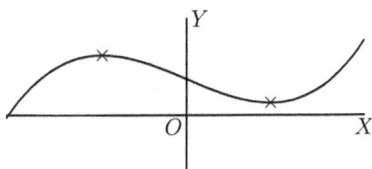

$$q \geq 2l\sqrt{l}$$

FIG. 18

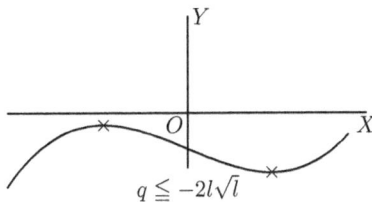

$$q \leq -2l\sqrt{l}$$

FIG. 19

If the equality sign holds in Fig. 18 or Fig. 19, a necessary and sufficient condition for which is $q^2 = 4l^3$, one of the bend points is on the x-axis, and the cubic equation has a double root. The inequalities in Fig. 20 hold if and only if $q^2 < 4l^3$, which implies that $l > 0$. Hence $x^3 - 3lx + q = 0$ *has three distinct real roots if and only if $q^2 < 4l^3$, a single real root if and only if $q^2 > 4l^3$, a double root (necessarily real) if and only if $q^2 = 4l^3$ and $l \neq 0$, and a triple root if $q^2 = 4l^3 = 0$.*

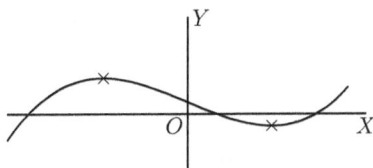

$$-2l\sqrt{l} < q < 2l\sqrt{l}$$

FIG. 20

EXERCISES

Find the bend points, sketch the graph, and find the number of real roots of

1. $x^3 + 2x - 4 = 0$.

3. $x^3 - 2x - 1 = 0$.

2. $x^3 - 7x + 7 = 0$.

4. $x^3 + 6x^2 - 3x + 1 = 0$.

5. Prove that the inflexion point of $y = x^3 - 3lx + q$ is $(0, q)$.

6. Show that the theorem in the text is equivalent to that in §45.

7. Prove that, if m and n are positive odd integers and $m > n$, $x^m + px^n + q = 0$ has no bend point and hence has a single real root if $p > 0$; but, if $p < 0$, it has just two bend points which are on the same side or opposite sides of the x-axis according as

$$\left(\frac{np}{m}\right)^m + \left(\frac{nq}{m-n}\right)^{m-n}$$

is positive or negative, so that the number of real roots is 1 or 3 in the respective cases.

8. Draw the graph of $y = x^4 - x^2$. By finding its intersections with the line $y = mx + b$, solve $x^4 - x^2 - mx - b = 0$.

9. Prove that, if p and q are positive, $x^{2m} - px^{2n} + q = 0$ has four distinct real roots, two pairs of equal roots, or no real root, according as

$$\left(\frac{np}{m}\right)^m - \left(\frac{nq}{m-n}\right)^{m-n} > 0, \quad = 0, \quad \text{or} \quad < 0.$$

10. Prove that no straight line crosses the graph of $y = f(x)$ in more than n points if the degree n of the real polynomial $f(x)$ exceeds unity. [Apply §16.] This fact serves as a check on the accuracy of a graph.

61. Definition of Continuity of a Polynomial.
Hitherto we have located certain points of the graph of $y = f(x)$, where $f(x)$ is a polynomial in x with real coefficients, and taken the liberty to join them by a continuous curve.

A polynomial $f(x)$ with real coefficients shall be called *continuous* at $x = a$, where a is a real constant, if the difference

$$D = f(a + h) - f(a)$$

is numerically less than any assigned positive number p for all real values of h sufficiently small numerically.

62. Any Polynomial $f(x)$ with real Coefficients is continuous at $x = a$, where a is any real Constant. Taylor's formula (8) gives

$$D = f'(a)h + \frac{f''(a)}{1 \cdot 2}h^2 + \cdots + \frac{f^{(n)}(a)}{1 \cdot 2 \cdots n}h^n.$$

This polynomial is a special case of

$$F = a_1 h + a_2 h^2 + \cdots + a_n h^n.$$

We shall prove that, *if $a_1, \ldots, a_n$ are all real, F is numerically less than any assigned positive number p for all real values of h sufficiently small numerically.* Denote by g the greatest numerical value of $a_1, \ldots, a_n$. If h is numerically less than k, where $k < 1$, we see that F is numerically less than

$$g(k + k^2 + \cdots + k^n) < g\frac{k}{1-k} < p, \qquad \text{if } k < \frac{p}{p+g}.$$

Hence a real polynomial $f(x)$ is continuous at every real value of x. But the function $\tan x$ is not continuous at $x = 90°$ (§63).

63. Root between a and b if $f(a)$ and $f(b)$ have opposite Signs. *If the coefficients of a polynomial $f(x)$ are real and if a and b are real numbers such that $f(a)$ and $f(b)$ have opposite signs, the equation $f(x) = 0$ has at least one real root between a and b; in fact, an odd number of such roots, if an m-fold root is counted as m roots.*

The only argument[2] given here (other than that in Ex. 5 below) is one based upon geometrical intuition. We are stating that, if the points

$$\left(a, f(a)\right), \qquad \left(b, f(b)\right)$$

lie on opposite sides of the x-axis, the graph of $y = f(x)$ crosses the x-axis once, or an odd number of times, between the vertical lines through these two points. Indeed, the part of the graph between these verticals is a continuous curve having one and only one point on each intermediate vertical line, since the function has a single value for each value of x.

This would not follow for the graph of $y^2 = x$, which is a parabola with the x-axis as its axis. It may not cross

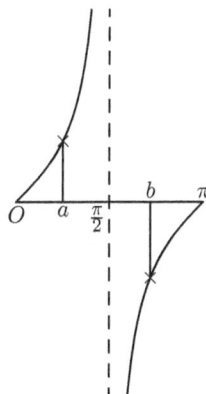

FIG. 21

---

[2]An arithmetical proof based upon a refined theory of irrational numbers is given in Weber's *Lehrbuch der Algebra*, ed. 2, vol. 1, p. 123.

the x-axis between the two initial vertical lines, but cross at a point to the left of each.

A like theorem does not hold for $f(x) = \tan x$, when x is measured in radians and $0 < a < \pi/2 < b < \pi$, since $\tan x$ is not continuous at $x = \pi/2$. When t increases from a to $\pi/2$, $\tan x$ increases without limit. When x decreases from b to $\pi/2$, $\tan x$ decreases without limit. There is no root between a and b of $\tan x = 0$.

EXERCISES

1. Prove that $8x^3 - 4x^2 - 18x + 9 = 0$ has a root between 0 and 1, one between 1 and 2, and one between -2 and -1.

2. Prove that $16x^4 - 24x^2 + 16x - 3 = 0$ has a triple root between 0 and 1, and a simple root between -2 and -1.

3. Prove that if $a < b < c \cdots < l$, and $\alpha, \beta, \ldots, \lambda$ are positive, these quantities being all real,

$$\frac{\alpha}{x-a} + \frac{\beta}{x-b} + \frac{\gamma}{x-c} + \cdots + \frac{\lambda}{x-l} + t = 0$$

has a real root between a and b, one between b and $c, \ldots$ one between k and l, and if t is negative one greater than l, but if t is positive one less than a.

4. Verify that the equation in Ex. 3 has no imaginary root by substituting $r + si$ and $r - si$ in turn for x, and subtracting the results.

5. Admitting that an equation $f(x) \equiv x^n + \cdots = 0$ with real coefficients has n roots, show algebraically that there is a real root between a and b if $f(a)$ and $f(b)$ have opposite signs. Note that a pair of conjugate imaginary roots $c \pm di$ are the roots of

$$(x - c)^2 + d^2 = 0$$

and that this quadratic function is positive if x is real. Hence if $x_1, \ldots, x_r$ are the real roots and

$$\phi(x) \equiv (x - x_1) \cdots (x - x_r),$$

then $\phi(a)$ and $\phi(b)$ have opposite signs. Thus $a - x_i$ and $b - x_i$ have opposite signs for at least one real root x_i. (Lagrange.)

64. Sign of a Polynomial.   Given a polynomial

$$f(x) = a_0 x^n + a_1 x^{n-1} + \cdots + a_n \qquad\qquad (a_0 \neq 0)$$

with real coefficients, we can find a positive number P such that $f(x)$ has the same sign as $a_0 x^n$ when $x > P$. In fact,

$$f(x) = x^n (a_0 + \phi), \qquad \phi = \frac{a_1}{x} + \frac{a_2}{x^2} + \cdots + \frac{a_n}{x^n}.$$

By the result in §62, the numerical value of ϕ is less than that of a_0 when $1/x$ is positive and less than a sufficiently small positive number, say $1/P$, and hence when $x > P$. Then $a_0 + \phi$ has the same sign as a_0, and hence $f(x)$ the same sign as $a_0 x^n$.

The last result holds also when x is a negative number sufficiently large numerically. For, if we set $x = -X$, the former case shows that $f(-X)$ has the same sign as $(-1)^n a_0 X^n$ when X is a sufficiently large positive number.

We shall therefore say briefly that, for $x = +\infty$, $f(x)$ has the same sign as a_0; while, for $x = -\infty$, $f(x)$ has the same sign as a_0 if n is even, but the sign opposite to a_0 if n is odd.

EXERCISES

1. Prove that $x^3 + ax^2 + bx - 4 = 0$ has a positive real root [use $x = 0$ and $x = +\infty$].

2. Prove that $x^3 + ax^2 + bx + 4 = 0$ has a negative real root [use $x = 0$ and $x = -\infty$].

3. Prove that if $a_0 > 0$ and n is odd, $a_0 x^n + \cdots + a_n = 0$ has a real root of sign opposite to the sign of a_n [use $x = -\infty$, 0, $+\infty$].

4. Prove that $x^4 + ax^3 + bx^2 + cx - 4 = 0$ has a positive and a negative root.

5. Show that any equation of even degree n in which the coefficient of x^n and the constant term are of opposite signs has a positive and a negative root.

65. Rolle's Theorem.   *Between two consecutive real roots a and b of $f(x) = 0$, there is an odd number of real roots of $f'(x) = 0$, a root of multiplicity m being counted as m roots.*

Let

$$f(x) \equiv (x-a)^r (x-b)^s Q(x), \qquad\qquad a < b,$$

where $Q(x)$ is a polynomial divisible by neither $x - a$ nor $x - b$. Then by the rule for the derivative of a product (§56, Ex. 10),

$$\frac{(x - a)(x - b)f'(x)}{f(x)} \equiv r(x - b) + s(x - a) + (x - a)(x - b)\frac{Q'(x)}{Q(x)}.$$

The second member has the value $r(a-b) < 0$ for $x = a$ and the value $s(b-a) > 0$ for $x = b$, and hence vanishes an odd number of times between a and b (§63). But, in the left member, $(x-a)(x-b)$ and $f(x)$ remain of constant sign between a and b, since $f(x) = 0$ has no root between a and b. Hence $f'(x)$ vanishes an odd number of times.

COROLLARY.   *Between two consecutive real roots α and β of $f'(x) = 0$ there occurs at most one real root of $f(x) = 0$.*

For, if there were two such real roots a and b of $f(x) = 0$, the theorem shows that $f'(x) = 0$ would have a real root between a and b and hence between α and β, contrary to hypothesis.

Applying also §63 we obtain the

CRITERION.   *If α and β are consecutive real roots of $f'(x) = 0$, then $f(x) = 0$ has a single real root between α and β if $f(\alpha)$ and $f(\beta)$ have opposite signs, but no root if they have like signs. At most one real root of $f(x) = 0$ is greater than the greatest real root of $f'(x) = 0$, and at most one real root of $f(x) = 0$ is less than the least real root of $f'(x) = 0$.*

If $f(\alpha) = 0$ for our root α of $f'(x) = 0, \alpha$ is a multiple root of $f(x) = 0$ and it would be removed before the criterion is applied.

EXAMPLE.   For $f(x) = 3x^5 - 25x^3 + 60x - 20$,

$$\tfrac{1}{15}f'(x) = x^4 - 5x^2 + 4 = (x^2 - 1)(x^2 - 4).$$

Hence the roots of $f'(x) = 0$ are $\pm 1, \pm 2$. Now

$$f(-\infty) = -\infty, \ f(-2) = -36, \ f(-1) = -58, \ f(1) = 18, \ f(2) = -4, \ f(+\infty) = +\infty.$$

Hence there is a single real root in each of the intervals

$$(-1, 1), \quad (1, 2), \quad (2, +\infty),$$

and two imaginary roots. The three real roots are positive.

EXERCISES

1. Prove that $x^5 - 5x + 2 = 0$ has 1 negative, 2 positive and 2 imaginary roots.

2. Prove that $x^6 + x - 1 = 0$ has 1 negative, 1 positive and 4 imaginary roots.

3. Show that $x^5 - 3x^3 + 2x^2 - 5 = 0$ has two imaginary roots, and a real root in each of the intervals $(-2, -1.5)$, $(-1.5, -1)$, $(1, 2)$.

4. Prove that $4x^5 - 3x^4 - 2x^2 + 4x - 10 = 0$ has a single real root.

5. Show that, if $f^{(k)}(x) = 0$ has imaginary roots, $f(x) = 0$ has imaginary roots.

6. Derive Rolle's theorem from the fact that there is an odd number of bend points between a and b, the abscissa of each being a root of $f'(x) = 0$ of odd multiplicity, while the abscissa of an inflexion point with a horizontal tangent is a root of $f'(x) = 0$ of even multiplicity.

CHAPTER VI

ISOLATION OF THE REAL ROOTS OF A REAL EQUATION

66. Purpose and Methods of Isolating the Real Roots. In the next chapter we shall explain processes of computing the real roots of a given real equation to any assigned number of decimal places. Each such method requires some preliminary information concerning the root to be computed. For example, it would be sufficient to know that the root is between 4 and 5, provided there be no other root between the same limits. But in the contrary case, narrower limits are necessary, such as 4 and 4.3, with the further fact that only one root is between these new limits. Then that root is said to be *isolated*.

If an equation has a single positive root and a single negative root, the real roots are isolated, since there is a single root between $-\infty$ and 0, and a single one between 0 and $+\infty$. However, for the practical purpose of their computation, we shall need narrower limits, sufficient to fix the first significant figure of each root, for example -40 and -30, or 20 and 30.

We may isolate the real roots of $f(x) = 0$ by means of the graph of $y = f(x)$. But to obtain a reliable graph, we saw in Chapter V that we must employ the bend points, whose abscissas occur among the roots of $f'(x) = 0$. Since the latter equation is of degree $n-1$ when $f(x) = 0$ is of degree n, this method is usually impracticable when n exceeds 3. The method based on Rolle's theorem (§65) is open to the same objection.

The most effective method is that due to Sturm (§68). We shall, however, begin with Descartes' rule of signs since it is so easily applied. Unfortunately it rarely tells us the exact number of real roots.

67. Descartes' Rule of Signs. Two consecutive terms of a real polynomial or equation are said to present a *variation of sign* if their coefficients have unlike signs. By the variations of sign of a real polynomial or equation we mean all the variations presented by consecutive terms.

Thus, in $x^5 - 2x^3 - 4x^2 + 3 = 0$, the first two terms present a variation of sign, and likewise the last two terms. The number of variations of sign of the equation is two.

DESCARTES' RULE   *The number of positive real roots of an equation with real coefficients is either equal to the number of its variations of sign or is less than that number by a positive even integer. A root of multiplicity m is here counted as m roots.*

For example, $x^6 - 3x^2 + x + 1 = 0$ has either two or no positive roots, the exact number not being found. But $3x^3 - x - 1 = 0$ has exactly one positive root, which is a simple root.

Descartes' rule will be derived in §73 as a corollary to Budan's theorem. The following elementary proof[1] was communicated to the author by Professor D. R. Curtiss.

Consider any real polynomial

$$f(x) \equiv a_0 x^n + a_1 x^{n-1} + \cdots + a_l x^{n-l} \qquad (a_0 \neq 0, \ a_l \neq 0).$$

Let r be a positive real number. By actual multiplication,

$$F(x) \equiv (x - r)f(x) \equiv A_0 x^{n+1} + A_1 x^n + \cdots + A_{l+1} x^{n-l},$$

where

$$A_0 = a_0, \quad A_1 = a_1 - ra_0, \quad A_2 = a_2 - ra_1, \ldots, A_l = a_l - ra_{l-1}, \quad A_{l+1} = -ra_l.$$

In $f(x)$ let a_{k_1} be the first non-vanishing coefficient of different sign from a_0, let a_{k_2} be the first non-vanishing coefficient following a_{k_1} and of the same sign as a_0, etc., the last such term, a_{k_v}, being either a_l or of the same sign as a_l. Evidently v is the number of variations of sign of $f(x)$.

For example, if $f(x) \equiv 2x^6 + 3x^5 - 4x^4 - 6x^3 + 7x$, we have $v = 2$, $a_{k_1} = a_2 = -4$, $a_{k_2} = a_5 = 7$. Note that $a_4 = 0$ since x^2 is absent.

The numbers $A_0, A_{k_1}, \ldots, A_{k_v}, A_{l+1}$ are all different from zero and have the same signs as $a_0, a_{k_1}, \ldots, a_{k_v}, -a_l$, respectively. This is obviously true for $A_0 = a_0$ and $A_{l+1} = -ra_l$. Next, A_{k_i} is the sum of the non-vanishing number a_{k_i} and the number $-ra_{k_i-1}$, which is either zero or else of the same sign as a_{k_i} since a_{k_i-1} is either zero or of opposite sign to a_{k_i}. Hence the sum A_{k_i} is not zero and has the same sign as a_{k_i}.

By hypothesis, each of the numbers $a_0, a_{k_1}, \ldots, a_{k_v}$ after the first is of opposite sign to its predecessor, while $-a_l$ is of opposite sign to a_{k_v}. Hence each term after the first in the sequence $A_0, A_{k_1}, \ldots, A_{k_v}, A_{l+1}$ is of opposite sign to its predecessor. Thus these terms present $v+1$ variations of sign. We conclude that $F(x)$ has at least one more variation of sign than $f(x)$. But we may go further and prove the following

---

[1]The proofs given in college algebras are mere verifications of special cases.

LEMMA.   *The number of variations of sign of $F(x)$ is equal to that of $f(x)$ increased by some positive odd integer.*

For, the sequence $A_0, A_1, \ldots, A_{k_1}$ has an odd number of variations of sign since its first and last terms are of opposite sign; and similarly for the v sequences

$$A_{k_1}, \quad A_{k_1+1}, \ldots, A_{k_2};$$
$$\cdots \cdots \cdots \cdots \cdots \cdots$$
$$A_{k_v}, \quad A_{k_v+1}, \ldots, A_{l+1}.$$

The total number of variations of sign of the entire sequence $A_0, A_1, \ldots, A_{l+1}$ is evidently the sum of the numbers of variations of sign for the $v + 1$ partial sequences indicated above, and is thus the sum of $v + 1$ positive odd integers. Since each such odd integer may be expressed as 1 plus 0 or a positive even integer, the sum mentioned is equal to $v + 1$ plus 0 or a positive even integer, i.e., to v plus a positive odd integer.

To prove Descartes' rule of signs, consider first the case in which $f(x) = 0$ has no positive real roots, i.e., no real root between 0 and $+\infty$. Then $f(0)$ and $f(\infty)$ are of the same sign (§63), and hence the first and last coefficients of $f(x)$ are of the same sign.[2] Thus $f(x)$ has either no variations of sign or an even number of them, as Descartes' rule requires.

Next, let $f(x) = 0$ have the positive real roots $r_1, \ldots, r_k$ and no others. A root of multiplicity m occurs here m times, so that the r's need not be distinct. Then

$$f(x) \equiv (x - r_1) \cdots (x - r_k)\phi(x),$$

where $\phi(x)$ is a polynomial with real coefficients such that $\phi(x) = 0$ has no positive real roots. We saw in the preceding paragraph that $\phi(x)$ has either no variations of sign or an even number of them. By the Lemma, the product $(x - r_k)\phi(x)$ has as the number of its variations of sign the number for $\phi(x)$ increased by a positive odd integer. Similarly when we introduce each new factor $x - r_i$. Hence the number of variations of sign of the final product $f(x)$ is equal to that of $\phi(x)$ increased by k positive odd integers, i.e., by k plus 0 or a positive even integer. Since $\phi(x)$ has either no variations of sign or an even number of them, the number of variations of sign of $f(x)$ is k plus 0 or a positive even integer, a result equivalent to our statement of Descartes' rule.

If $-p$ is a negative root of $f(x) = 0$, then p is a positive root of $f(-x) = 0$. Hence we obtain the

COROLLARY.   *The number of* negative *roots of $f(x) = 0$ is either equal to the number of variations of sign of $f(-x)$ or is less than that number by a positive even integer.*

---

[2] In case $f(x)$ has a factor x^{n-l}, we use the polynomial $f(x)/x^{n-l}$ instead of $f(x)$ in this argument.

For example, $x^4 + 3x^3 + x - 1 = 0$ has a single negative root, which is a simple root, since $x^4 - 3x^3 - x - 1 = 0$ has a single positive root.

As indicated in Exs. 10, 11 below, Descartes' rule may be used to isolate the roots.

EXERCISES

Prove by Descartes' rule the statements in Exs. 1–8, 12, 15.

1. An equation all of whose coefficients are of like sign has no positive root. Why is this self-evident?

2. There is no negative root of an equation, like $x^5 - 2x^4 - 3x^2 + 7x - 5 = 0$, in which the coefficients of the odd powers of x are of like sign, and the coefficients of the even powers (including the constant term) are of the opposite sign. Verify by taking $x = -p$, where p is positive.

3. $x^3 + a^2x + b^2 = 0$ has two imaginary roots if $b \neq 0$.

4. For n even, $x^n - 1 = 0$ has only two real roots.

5. For n odd, $x^n - 1 = 0$ has only one real root.

6. For n even, $x^n + 1 = 0$ has no real root; for n odd, only one.

7. $x^4 + 12x^2 + 5x - 9 = 0$ has just two imaginary roots.

8. $x^4 + a^2x^2 + b^2x - c^2 = 0$ $(c \neq 0)$ has just two imaginary roots.

9. Descartes' rule enables us to find the exact number of positive roots only when all the coefficients are of like sign or when

$$f(x) = x^n + p_1 x^{n-1} + \cdots + p_{n-s} x^s - p_{n-s+1} x^{s-1} - \cdots - p_n = 0,$$

each p_i being ≥ 0. Without using that rule, show that the latter equation has one and only one positive root r. Hints: There is a positive root r by §63 $(a = 0, b = \infty)$. Denote by $P(x)$ the quotient of the sum of the positive terms by x^s, and by $-N(x)$ that of the negative terms. Then $N(x)$ is a sum of powers of $1/x$ with positive coefficients.

If $x > r$, $P(x) > P(r)$, $N(x) < N(r)$, $f(x) > 0$;

If $x < r$, $P(x) < P(r)$, $N(x) > N(r)$, $f(x) < 0$.

10. Prove that we obtain an upper limit to the number of real roots of $f(x) = 0$ between a and b, if we set

$$x = \frac{a + by}{1 + y} \qquad \left(\therefore y = \frac{x - a}{b - x} \right),$$

multiply by $(1 + y)^n$, and apply Descartes' rule to the resulting equation in y.

11. Show by the method of Ex. 10 that there is a single root between 2 and 4 of $x^3 + x^2 - 17x + 15 = 0$. Here we have $27y^3 + 3y^2 - 23y - 7 = 0$.

12. In the astronomical problem of three bodies occurs the equation

$$r^5 + (3 - \mu)r^4 + (3 - 2\mu)r^3 - \mu r^2 - 2\mu r - \mu = 0,$$

where $0 < \mu < 1$. Why is there a single positive real root?

13. Prove that $x^5 + x^3 - x^2 + 2x - 3 = 0$ has four imaginary roots by applying Descartes' rule to the equation in y whose roots are the squares of the roots of the former. Transpose the odd powers, square each new member, and replace x^2 by y.

14. As in Ex. 13 prove that $x^3 + x^2 + 8x + 6 = 0$ has imaginary roots.

15. If a real equation $f(x) = 0$ of degree n has n real roots, the number of positive roots is exactly equal to the number V of variations of sign. Hint: consider also $f(-x)$.

16. Show that $x^3 - x^2 + 2x + 1 = 0$ has no positive root. Hint: multiply by $x + 1$.

68. Sturm's Method. Let $f(x) = 0$ be an equation with real coefficients, and $f'(x)$ the first derivative of $f(x)$. The first step of the usual process of finding the greatest common divisor of $f(x)$ and $f'(x)$, if it exists, consists in dividing f by f' until we obtain a remainder $r(x)$, whose degree is less than that of f'. Then, if q_1 is the quotient, we have $f = q_1 f' + r$. Instead of dividing f' by r, as in the greatest common divisor process, and proceeding further in that manner, we write $f_2 = -r$, divide f' by f_2, and denote by f_3 the remainder with its sign changed. Thus

$$f = q_1 f' - f_2, \qquad f' = q_2 f_2 - f_3, \qquad f_2 = q_3 f_3 - f_4, \dots .$$

The latter equations, in which each remainder is exhibited as the negative of a polynomial f_i, yield a modified process, just as effective as the usual process, of finding the greatest common divisor G of $f(x)$ and $f'(x)$ if it exists.

Suppose that $-f_4$ is the first constant remainder. If $f_4 = 0$, then $f_3 = G$, since f_3 divides f_2 and hence also f' and f (as shown by using our above equations in reverse order); while, conversely, any common divisor of f and f' divides f_2 and hence also f_3.

But if f_4 is a constant $\neq 0$, f and f' have no common divisor involving x. This case arises if and only if $f(x) = 0$ has no multiple root (§58), and is the only case considered in §§69–71.

Before stating Sturm's theorem in general, we shall state it for a numerical case and illustrate its use.

EXAMPLE. $f(x) = x^3 + 4x^2 - 7$. Then $f' = 3x^2 + 8x$,

$$f = (\tfrac{1}{3}x + \tfrac{4}{9})f' - f_2, \qquad f_2 \equiv \tfrac{32}{9}x + 7,$$
$$f' = (\tfrac{27}{32}x + \tfrac{603}{1024})f_2 - f_3, \qquad f_3 = \tfrac{4221}{1024}.$$

For[3] $x = 1$, the signs of f, f', f_2, f_3, are $- + + +$, showing a single variation of consecutive signs. For $x = 2$, the signs are $+ + + +$, showing no variation of sign. Sturm's theorem states that there is a *single* real root between 1 and 2. For $x = -\infty$, the signs are $- + - +$, showing 3 variations of sign. The theorem states that there are $3 - 1 = 2$ real roots between $-\infty$ and 1. Similarly,

x	Signs	Variations
-1	$- - + +$	1
-2	$+ - - +$	2
-3	$+ + - +$	2
-4	$- + - +$	3

Hence there is a single real root between -2 and -1, and a single one between -4 and -3. Each real root has now been *isolated* since we have found two numbers such that a single real root lies between these two numbers or is equal to one of them.

Some of the preceding computation was unnecessary. After isolating a root between -2 and -1, we know that the remaining root is isolated between $-\infty$ and -2. But before we can compute it by Horner's method, we need closer limits for it. For that purpose it is unnecessary to find the signs of all four functions, but merely the sign of f (§63).

69. Sturm's Theorem. *Let $f(x) = 0$ be an equation with real coefficients and without multiple roots. Modify the usual process of seeking the greatest common divisor of $f(x)$ and its first derivative[4] $f_1(x)$ by exhibiting each remainder as the negative of a polynomial f_i:*

(1) $f = q_1 f_1 - f_2, \; f_1 = q_2 f_2 - f_3, \; f_2 = q_3 f_3 - f_4, \ldots, \; f_{n-2} = q_{n-1} f_{n-1} - f_n,$

where[5] f_n is a constant $\neq 0$. If a and b are real numbers, $a < b$, neither a root of $f(x) = 0$, the number of real roots of $f(x) = 0$ between a and b is equal to the excess of the number of variations of sign of

(2) $f(x), \quad f_1(x), \quad f_2(x), \ldots, f_{n-1}(x), \quad f_n$

---

[3]Before going further, check that the preceding relations hold when $x = 1$ by inserting the computed values of f, f', f_2 for $x = 1$. Experience shows that most students make some error in finding $f_2, f_3, \ldots$, so that checking is essential.

[4]The notation f_1 instead of the usual f', and similarly f_0 instead of f, is used to regularize the notation of all the f's, and enables us to write any one of the equations (1) in the single notation (3).

[5]If the division process did not yield ultimately a constant remainder $\neq 0$, f and f_1 would have a common factor involving x, and hence $f(x) = 0$ a multiple root.

for $x = a$ *over the number of variations of sign for* $x = b$. *Terms which vanish are to be dropped out before counting the variations of sign.*

For brevity, let V_x denote the number of variations of sign of the numbers (2) when x is a particular real number not a root of $f(x) = 0$.

First, if x_1 and x_2 are real numbers such that no one of the continuous functions (2) vanishes for a value of x between x_1 and x_2 or for $x = x_1$ or $x = x_2$, the values of any one of these functions for $x = x_1$ and $x = x_2$ are both positive or both negative (§63), and therefore $V_{x_1} = V_{x_2}$.

Second, let ρ be a root of $f_i(x) = 0$, where $1 \leq i < n$. Then

$$(3) \qquad\qquad f_{i-1}(x) = q_i f_i(x) - f_{i+1}(x)$$

and the equations (1) following this one show that $f_{i-1}(x)$ and $f_i(x)$ have no common divisor involving x (since it would divide the constant f_n). By hypothesis, $f_i(x)$ has the factor $x - \rho$. Hence $f_{i-1}(x)$ does not have this factor $x - \rho$. Thus, by (3),

$$f_{i-1}(\rho) = -f_{i+1}(\rho) \neq 0.$$

Hence, if p is a sufficiently small positive number, the values of

$$f_{i-1}(x), \quad f_i(x), \quad f_{i+1}(x)$$

for $x = \rho - p$ show just one variation of sign, since the first and third values are of opposite sign, and for $x = \rho + p$ show just one variation of sign, and therefore show no change in the number of variations of sign for the two values of x.

It follows from the first and second cases that $V_\alpha = V_\beta$ if α and β are real numbers for neither of which any one of the functions (2) vanishes and such that no root of $f(x) = 0$ lies between α and β.

Third, let r be a root of $f(x) = 0$. By Taylor's theorem (8) of §56,

$$f(r - p) = -pf'(r) + \tfrac{1}{2}p^2 f''(r) - \cdots,$$
$$f(r + p) = pf'(r) + \tfrac{1}{2}p^2 f''(r) + \cdots.$$

If p is a sufficiently small positive number, each of these polynomials in p has the same sign as its first term. For, after removing the factor p, we obtain a quotient of the form $a_0 + s$, where $s = a_1 p + a_2 p^2 + \cdots$ is numerically less than a_0 for all values of p sufficiently small (§62). Hence if $f'(r)$ is positive, $f(r - p)$ is negative and $f(r + p)$ is positive, so that the terms $f(x)$, $f_1(x) \equiv f'(x)$ have the signs $- +$ for $x = r - p$ and the signs $+ +$ for $x = r + p$. If $f'(r)$ is negative, these signs are $+ -$ and $- -$ respectively. In each case, $f(x)$, $f_1(x)$ show one more variation of sign for $x = r - p$ than for $x = r + p$. Evidently p may be chosen so small that no one of the functions $f_1(x), \ldots, f_n$ vanishes for either $x = r - p$ or $x = r + p$, and such that $f_1(x)$ does not vanish for a value of x between $r - p$ and $r + p$, so that $f(x) = 0$ has the single real root r between these limits (§65). Hence by the first and second cases, $f_1, \ldots, f_n$ show the same number of variations of sign for $x = r - p$ as for $x = r + p$. Thus, for the entire series of functions (2), we have

(4) $$V_{r-p} - V_{r+p} = 1.$$

The real roots of $f(x) = 0$ within the main interval from a to b (i.e., the aggregate of numbers between a and b) separate it into intervals. By the earlier result, V_x has the same value for all numbers in the same interval. By the present result (4), the value V_x in any interval exceeds the value for the next interval by unity. Hence V_a exceeds V_b by the number of real roots between a and b.

COROLLARY.   If $a < b$, then $V_a \geqq V_b$.

A violation of this Corollary usually indicates an error in the computation of Sturm's functions (2).

EXERCISES

Isolate by Sturm's theorem the real roots of

1. $x^3 + 2x + 20 = 0.$ **2.** $x^3 + x - 3 = 0.$

70. Simplifications of Sturm's Functions.   In order to avoid fractions, we may first multiply $f(x)$ by a *positive* constant before dividing it by by $f_1(x)$, and similarly multiply f_1 by a positive constant before dividing it by f_2, etc. Moreover, we may remove from any f_i any factor k_i which is either a positive constant or a polynomial in x positive for[6] $a \leqq x \leqq b$, and use the remaining factor F_i as the next divisor.

To prove that Sturm's theorem remains true when these modified functions $f, F_1, \ldots, F_m$ are employed in place of functions (2), consider the equations replacing (1):

$$f_1 = k_1 F_1, \qquad c_2 f = q_1 F_1 - k_2 F_2, \qquad c_3 F_1 = q_2 F_2 - k_3 F_3,$$
$$c_4 F_2 = q_3 F_3 - k_4 F_4, \ldots, c_m F_{m-2} = q_{m-1} F_{m-1} - k_m F_m,$$

---

[6]Usually we would require that k_i be positive for all values of x, since we usually wish to employ the limits $-\infty$ and $+\infty$.

in which $c_2, c_3, \ldots$ are positive constants and F_m is a constant $\neq 0$. A common divisor (involving x) of F_{i-1} and F_i would divide $F_{i-2}, \ldots, F_2, F_1, f, f_1$, whereas $f(x) = 0$ has no multiple roots. Hence if ρ is a root of $F_i(x) = 0$, then $F_{i-1}(\rho) \neq 0$ and

$$c_{i+1} F_{i-1}(\rho) = -k_{i+1}(\rho) F_{i+1}(\rho), \qquad c_{i+1} > 0, \qquad k_{i+1}(\rho) > 0.$$

Thus F_{i-1} and F_{i+1} have opposite signs for $x = \rho$. We proceed as in §69.

EXAMPLE 1.   If $f(x) = x^3 + 6x - 10$, $f_1 = 3(x^2 + 2)$ is always positive. Hence we may employ f and $F_1 = 1$. For $x = -\infty$, there is one variation of sign; for $x = +\infty$, no variation. Hence there is a single real root; it lies between 1 and 2.

EXAMPLE 2.   If $f(x) = 2x^4 - 13x^2 - 10x - 19$, we may take

$$f_1 = 4x^3 - 13x - 5.$$

Then

$$2f = xf_1 - f_2, \qquad f_2 = 13x^2 + 15x + 38 = 13(x + \tfrac{15}{26})^2 + \tfrac{1751}{52}.$$

Since f_2 is always positive, we need go no further (we may take $F_2 = 1$). For $x = -\infty$, the signs are $+ - +$; for $x = +\infty$, $+ + +$. Hence there are two real roots. The signs for $x = 0$ are $- - +$. Hence one real root is positive and the other negative.

EXERCISES

Isolate by Sturm's theorem the real roots of

1. $x^3 + 3x^2 - 2x - 5 = 0$.   2. $x^4 + 12x^2 + 5x - 9 = 0$.

3. $x^3 - 7x - 7 = 0$.   4. $3x^4 - 6x^2 + 8x - 3 = 0$.

5. $x^6 + 6x^5 - 30x^2 - 12x - 9 = 0$ [stop with f_2].

6. $x^4 - 8x^3 + 25x^2 - 36x + 8 = 0$.

7. For $f = x^3 + px + q$ $(p \neq 0)$, show that $f_1 = 3x^2 + p$, $f_2 = -2px - 3q$,

$$4p^2 f_1 = (-6px + 9q)f_2 - f_3, \qquad f_3 = -4p^3 - 27q^2,$$

so that f_3 is the discriminant Δ (§44). Let $[p]$ denote the sign of p. Then the signs of f, f_1, f_2, f_3 are

$$- + + [p] \; [\Delta] \quad \text{for } x = -\infty,$$
$$+ + - [p] \; [\Delta] \quad \text{for } x = +\infty.$$

For Δ negative there is a single real root. For Δ positive and therefore p negative, there are three distinct real roots. For $\Delta = 0$, f_2 is a divisor of f_1 and f, so that $x = -3q/(2p)$ is a double root.

8. Prove that if one of Sturm's functions has p imaginary roots, the initial equation has at least p imaginary roots.

9. State Sturm's theorem so as to include the possibility of a, or b, or both a and b being roots of $f(x) = 0$.

71. Sturm's Functions for a Quartic Equation. For the reduced quartic equation $f(z) = 0$,

(5)
$$\begin{cases} f = z^4 + qz^2 + rz + s, \\ f_1 = 4z^3 + 2qz + r, \\ f_2 = -2qz^2 - 3rz - 4s. \end{cases}$$

Let $q \neq 0$ and divide $q^2 f_1$ by f_2. The negative of the remainder is

(6)
$$f_3 = Lz - 12rs - rq^2, \qquad L = 8qs - 2q^3 - 9r^2.$$

Let $L \neq 0$. Then f_4 is a constant which is zero if and only if $f = 0$ has multiple roots, i.e., if its discriminant Δ is zero. We therefore desire f_4 expressed as a multiple of Δ. By §50,

(7)
$$\Delta = -4P^3 - 27Q^2, \qquad P = -4s - \frac{q^2}{3}, \qquad Q = \tfrac{8}{3}qs - r^2 - \tfrac{2}{27}q^3.$$

We may employ P and Q to eliminate

(8)
$$4s = -P - \frac{q^2}{3}, \qquad r^2 = -Q - \tfrac{2}{3}qP - \tfrac{8}{27}q^3.$$

We divide $L^2 f_2$ by

(9)
$$f_3 = Lz + 3rP, \qquad L = 9Q + 4qP.$$

The negative of the remainder[7] is

(10)
$$18r^2qP^2 - 9r^2LP + 4sL^2 = q^2\Delta.$$

The left member is easily reduced to $q^2\Delta$. Inserting the values (8) and replacing L^2 by $L(9Q + 4qP)$, we get

$$-18qQP^2 - 12q^2P^3 - \tfrac{16}{3}q^4P^2 + 2qP^2L + \tfrac{4}{3}q^3PL - 3q^2QL.$$

---

[7]Found directly by the Remainder Theorem (§14) by inserting the root $z = -3rP/L$ of $f_3 = 0$ into $L^2 f_2$.

Replacing L by its value (9), we get $q^2\Delta$. Hence we may take

$$(11) \qquad\qquad f_4 = \Delta.$$

Hence if $qL\Delta \neq 0$, we may take (5), (9), (11) as Sturm's functions.
Denote the sign of q by $[q]$. The signs of Sturm's functions are

$$
\begin{array}{cccccc}
+ & - & -[q] & -[L] & [\Delta] & \text{for } x = -\infty, \\
+ & + & -[q] & [L] & [\Delta] & \text{for } x = +\infty.
\end{array}
$$

First, let $\Delta > 0$. If q is negative and L is positive, the signs are $+ - + - +$
and $+ + + + +$, so that there are four real roots. In each of the remaining
three cases for q and L, there are two variations of sign in either of the two
series and hence there is no real root.

Next, let $\Delta < 0$. In each of the three cases in which q and L are not both
positive, there are three variations of sign in the first series and one variation
in the second, and hence just two real roots. If q and L are both positive, the
number of variations is 1 in the first series and 3 in the second, so that this
case is excluded by the Corollary to Sturm's theorem. To give a direct proof,
note that, by the value of L in (6), $L > 0$, $q > 0$ imply $4s > q^2$, i.e., $s > 0$, and
hence, by (7), P is negative, so that each term of (10) is ≥ 0, whence $\Delta > 0$.

Hence, if $qL\Delta \neq 0$, there are four distinct real roots if and only if Δ and L
are positive, and q negative; two distinct real and two imaginary roots if and
only if Δ is negative.

Combining this result with that in Ex. 4 below, we obtain the

THEOREM.   *If the discriminant Δ of $z^4 + qz^2 + rz + s = 0$ is negative, there
are two distinct real roots and two imaginary roots; if $\Delta > 0$, $q < 0$, $L > 0$,
four distinct real roots; if $\Delta > 0$ and either $q \geq 0$ or $L \leq 0$, no real roots. Here
$L = 8qs - 2q^3 - 9r^2$.*

Our discussion furnished also the series of Sturm functions, which may be
used in isolating the roots.

EXERCISES

1. If $q\Delta \neq 0$, $L = 0$, then $f_3 = 3rP$ is not zero (there being no multiple root)
and its sign is immaterial in determining the number of real roots. Prove that there
are just two real roots if $q < 0$, and none if $q > 0$. By (10), q has the same sign as Δ.

2. If $r\Delta \neq 0$, $q = 0$, obtain $-f_3$ by substituting $z = -4s/(3r)$ in f_1. Show that
we may take $f_3 = r\Delta$ and that there are just two real roots if $\Delta < 0$, and no real
roots if $\Delta > 0$.

3. If $\Delta \neq 0$, $q = r = 0$, prove that there are just two real roots if $\Delta < 0$, and no real roots if $\Delta > 0$. Since $\Delta = 256s^3$, check by solving $z^4 + s = 0$.

4. If $\Delta \neq 0$, $qL = 0$, there are just two real roots if $\Delta < 0$, and no real roots if $\Delta > 0$. [Combine the results in Exs. 1–3.]

5. Apply the theorem to Exs. 2, 4, 6 of §70.

6. Isolate the real roots of Exs. 3, 4, 5 of §48.

72. Sturm's Theorem for the Case of Multiple Roots. We might remove the multiple roots by dividing $f(x)$ by[8] $f_n(x)$, the greatest common divisor of $f(x)$ and $f_1 = f'(x)$; but this would involve considerable work, besides wasting the valuable information in hand. As before, we suppose $f(a)$ and $f(b)$ different from zero. We have equations (1) in which f_n is now not a constant.

The difference $V_a - V_b$ is the number of real roots between a and b, each multiple root being counted only once.

If ρ is a root of $f_i(x) = 0$, but not a multiple root of $f(x) = 0$, then $f_{i-1}(\rho)$ is not zero. For, if it were zero, $x - \rho$ would by (1) be a common factor of f and f_1. We may now proceed as in the second case in §69.

The third case requires a modified proof only when r is a multiple root. Let r be a root of multiplicity m, $m \geq 2$. Then $f(r)$, $f'(r), \ldots, f^{(m-1)}(r)$ are zero and, by Taylor's theorem,

$$f(r + p) = \frac{p^m}{1 \cdot 2 \cdots m} f^{(m)}(r) + \cdots ,$$

$$f'(r + p) = \frac{p^{m-1}}{1 \cdot 2 \cdots (m - 1)} f^{(m)}(r) + \cdots .$$

These have like signs if p is a positive number so small that the signs of the polynomials are those of their first terms. Similarly, $f(r - p)$ and $f'(r - p)$ have opposite signs. Hence f and f_1 show one more variation of sign for $x = r - p$ than for $x = r + p$. Now $(x - r)^{m-1}$ is a factor of f and f_1 and hence, by (1), of $f_2, \ldots, f_n$. Let their quotients by this factor be $\phi, \phi_1, \ldots, \phi_n$. Then equations (1) hold after the f's are replaced by the ϕ's. Taking p so small that $\phi_1(x) = 0$ has no root between $r - p$ and $r + p$, we see by the first and second cases in §69 that $\phi_1, \ldots, \phi_n$ show the same number of variations of sign for $x = r - p$ as for $x = r + p$. The same is true for $f_1, \ldots, f_n$ since the products of $\phi_1, \ldots, \phi_n$ by $(x - r)^{m-1}$ have for a given x the same signs as $\phi_1, \ldots, \phi_n$ or the same signs as $-\phi_1, \ldots, -\phi_n$. But the latter series evidently shows the same number of variations of sign as $\phi_1, \ldots, \phi_n$. Hence (4) is proved and consequently the present theorem.

---

[8] The degree of $f(x)$ is not n, nor was it necessarily n in §69.

EXERCISES

1. For $f = x^4 - 8x^2 + 16$, prove that $F_1 = x^3 - 4x$, $F_2 = x^2 - 4$, $F_1 = xF_2$. Hence $n = 2$. Verify that $V_{-\infty} = 2$, $V_{\infty} = 0$, and that there are just two real roots, each a double root.

Discuss similarly the following equations.

2. $x^4 - 5x^3 + 9x^2 - 7x + 2 = 0$.

3. $x^4 + 2x^3 - 3x^2 - 4x + 4 = 0$.

4. $x^4 - x^2 - 2x + 2 = 0$.

73. Budan's Theorem. *Let a and b be real numbers, $a < b$, neither[9] a root of $f(x) = 0$, an equation of degree n with real coefficients. Let V_a denote the number of variations of sign of*

$$(12) \qquad f(x), \qquad f'(x), \qquad f''(x), \qquad \ldots, \qquad f^{(n)}(x)$$

for $x = a$, after vanishing terms have been deleted. Then $V_a - V_b$ is either the number of real roots of $f(x) = 0$ between a and b or exceeds the number of those roots by a positive even integer. A root of multiplicity m is here counted as m roots.

For example, if $f(x) = x^3 - 7x - 7$, then $f' = 3x^2 - 7$, $f'' = 6x$, $f''' = 6$. Their values for $x = 3, 4, -2, -1$ are tabulated below.

x	f	f'	f''	f'''	Variations
3	-1	20	18	6	1
4	29	41	24	6	0
-2	-1	5	-12	6	3
-1	-1	-4	-6	6	1

Hence the theorem shows that there is a single real root between 3 and 4, and two or no real roots between -2 and -1. The theorem does not tell us the exact number of roots between the latter limits. To decide this ambiguity, note that $f(-3/2) = +1/8$, so that there is a single real root between -2 and -1.5, and a single one between -1.5 and -1.

The proof is quite simple if no term of the series (12) vanishes for $x = a$ or for $x = b$ and if no two consecutive terms vanish for the same value of x between a and b. Indeed, if no one of the terms vanishes for $x_1 \leqq x \leqq x_2$, then

---

[9] In case a or b is a root of $f(x) = 0$, the theorem holds if we count the number of roots $> a$ and $\leqq b$. This inclusive theorem has been proved, by means of Rolle's theorem, by A. Hurwitz, *Mathematische Annalen*, Vol. 71, 1912, p. 584, who extended Budan's theorem from the case of a polynomial to a function $f(x)$ which is real and regular for $a \leqq x < b$.

$V_{x_1} = V_{x_2}$, since any term has the same sign for $x = x_1$ as for $x = x_2$. Next, let r be a root of $f^{(i)}(x) = 0$, $a < r < b$. By hypothesis, the first derivative $f^{(i+1)}(x)$ of $f^{(i)}(x)$ is not zero for $x = r$. As in the third step (now actually the case $i = 0$) in §69, $f^{(i)}(x)$ and $f^{(i+1)}(x)$ show one more variation of sign for $x = r - p$ than for $x = r + p$, where p is a sufficiently small positive number. If $i > 0$, $f^{(i)}$ is preceded by a term $f^{(i-1)}$ in (12). By hypothesis, $f^{(i-1)}(x) \neq 0$ for $x = r$ and hence has the same sign for $x = r - p$ and $x = r + p$ when p is sufficiently small. For these values of x, $f^{(i)}(x)$ has opposite signs. Hence $f^{(i-1)}$ and $f^{(i)}$ show one more or one less variation of sign for $x = r - p$ than for $x = r + p$, so that $f^{(i-1)}$, $f^{(i)}$, $f^{(i+1)}$ show two more variations or the same number of variations of sign.

Next, let no term of the series (12) vanish for $x = a$ or for $x = b$, but let several successive terms

$$(13) \qquad\qquad f^{(i)}(x), \qquad f^{(i+1)}(x), \ldots, f^{(i+j-1)}(x)$$

all vanish for a value r of x between a and b, while $f^{(i+j)}(r)$ is not zero, but is say positive.[10] Let I_1 be the interval between $r - p$ and r, and I_2 the interval between r and $r + p$. Let the positive number p be so small that no one of the functions (13) or $f^{(i+j)}(x)$ is zero in these intervals, so that the last function remains positive. Hence $f^{(i+j-1)}(x)$ increases with x (since its derivative is positive) and is therefore negative in I_1 and positive in I_2. Thus $f^{(i+j-2)}(x)$ decreases in I_1 and increases in I_2 and hence is positive in each interval. In this manner we may verify the signs in the following table:

	$f^{(i)}$	$f^{(i+1)}$	$f^{(i+2)}$	$\ldots$	$f^{(i+j-3)}$	$f^{(i+j-2)}$	$f^{(i+j-1)}$	$f^{(i+j)}$
I_1	$(-)^j$	$(-)^{j-1}$	$(-)^{j-2}$	$\ldots$	$-$	$+$	$-$	$+$
I_2	$+$	$+$	$+$	$\ldots$	$+$	$+$	$+$	$+$

Hence these functions show j variations of sign in I_1 and none in I_2.

If $i > 0$, the first term of (13) is preceded by a function $f^{(i-1)}(x)$ which is not zero for $x = r$, and hence not zero in I_1 or I_2 if p is sufficiently small. If j is even, the signs of $f^{(i-1)}$ and $f^{(i)}$ are $+ +$ or $- +$ in both I_1 and I_2, showing no loss in the number of variations of sign. If j is odd, their signs are

I_1	$+ -$		$- -$
		or	
I_2	$+ +$		$- +$

so that there is a loss or gain of a single variation of sign. Hence

$$f^{(i-1)}, \qquad f^{(i)}, \qquad f^{(i+1)} \quad \ldots, \qquad f^{(i+j)}$$

---

[10] If negative, all signs in the table below are to be changed; but the conclusion holds.

show a loss of j variations of sign if j is even, and a loss of $j \pm 1$ if j is odd, and hence always a loss of an even number ≥ 0 of variations of sign.

If $i = 0$, $f^{(i)} \equiv f$ has r as a j-fold root and the functions in the table show j more variations of sign for $x = r - p$ than for $x = r + p$.

Thus, when no one of the functions (12) vanishes for $x = a$ or for $x = b$, the theorem follows as at the end of §69 (with unity replaced by the multiplicity of a root).

Finally, let one of the functions (12), other than $f(x)$ itself, vanish for $x = a$ or for $x = b$. If δ is a sufficiently small positive number, all of the N roots of $f(x) = 0$ between a and b lie between $a + \delta$ and $b - \delta$, and for the latter values no one of the functions (12) is zero. By the above proof,

$$V_{a+\delta} - V_{b-\delta} = N + 2t,$$

$$V_a - V_{a+\delta} = 2j, \qquad V_{b-\delta} - V_b = 2s,$$

where t, j, s are integers ≥ 0. Hence $V_a - V_b = N + 2(t + j + s)$.

Descartes' rule of signs (§67) is a corollary to Budan's theorem. Consider any equation with real coefficients

$$f(x) \equiv a_0 x^n + a_1 x^{n-1} + \cdots + a_{n-1} x + a_n = 0,$$

having $a_n \neq 0$. For $x = 0$ the functions (12) have the same signs as

$$a_n, \qquad a_{n-1}, \qquad \ldots, \qquad a_1, \qquad a_0.$$

Hence V_0 is equal to the number V of variations of sign of $f(x)$.

For $x = +\infty$, the functions all have the same sign, which is that of a_0. Thus $V_0 - V_\infty = V$ is either the number of positive roots or exceeds that number by a positive even integer. Finally, Descartes' rule holds if $a_n = 0$, as shown by removing the factors x.

EXERCISES

Isolate by Budan's theorem the real roots of

1. $x^3 - x^2 - 2x + 1 = 0$.   **2.** $x^3 + 3x^2 - 2x - 5 = 0$.

3. Prove that if $f(a) \neq 0$, V_a equals the number of real roots $> a$ or exceeds that number by an even integer.

4. Prove that there is no root greater than a number making each of the functions (12) positive, if the leading coefficient of $f(x)$ is positive. (Newton.)

5. Hence verify that $x^4 - 4x^3 - 3x + 23 = 0$ has no root > 4.

6. Show that $x^4 - 4x^3 + x^2 + 6x + 2 = 0$ has no root > 3.

CHAPTER VII

SOLUTION OF NUMERICAL EQUATIONS

74. Horner's Method.[1]   After we have isolated a real root of a real equation by one of the methods in Chapter VI, we can compute the root to any desired number of decimal places either by Horner's method, which is available only for polynomial equations, or by Newton's method (§75), which is applicable also to logarithmic, trigonometric, and other equations.

To find the root between 2 and 3 of

$$(1) \qquad x^3 - 2x - 5 = 0,$$

set $x = 2 + p$. Direct substitution gives the *transformed equation* for p:

$$(2) \qquad p^3 + 6p^2 + 10p - 1 = 0.$$

The method just used is laborious especially for equations of high degree. We next explain a simpler method. Since $p = x - 2$,

$$x^3 - 2x - 5 \equiv (x - 2)^3 + 6(x - 2)^2 + 10(x - 2) - 1,$$

identically in x. Hence -1 is the remainder obtained when the given polynomial $x^3 - 2x - 5$ is divided by $x - 2$. By inspection, the quotient Q is equal to

$$(x - 2)^2 + 6(x - 2) + 10.$$

Hence 10 is the remainder obtained when Q is divided by $x - 2$. The new quotient is equal to $(x - 2) + 6$, and another division gives the remainder 6. Hence to find the coefficients 6, 10, -1 of the terms following p^3 in the transformed equation (2), we have only to divide the given polynomial $x^3 - 2x - 5$ by $x - 2$, the quotient Q by $x - 2$, etc., and take the remainders in reverse order. However, when this work is performed by synthetic division (§15) as

---

[1] W. G. Horner, London Philosophical Transactions, 1819. Earlier (1804) by P. Ruffini. See Bulletin American Math. Society, May, 1911.

tabulated below, no reversal of order is necessary, since the coefficients then appear on the page in their desired order.

$$
\begin{array}{ccccc}
1 & 0 & -2 & -5 & \big|\,2 \\
  & 2 & 4 & 4 & \\
\hline
1 & 2 & 2 & -1 & \\
  & 2 & 8 & & \\
\hline
1 & 4 & 10 & & \\
  & 2 & & & \\
\hline
1 & 6 & & & \\
\end{array}
$$

Thus 1, 6, 10, −1 are the coefficients of the desired equation (2).

To obtain an approximation to the decimal p, we ignore for the moment the terms involving p^3 and p^2; then by $10p - 1 = 0$, $p = 0.1$. But this value is too large since the terms ignored are all positive. For $p = 0.09$, the polynomial in (2) is found to be negative, while for $p = 0.1$ it was just seen to be positive. Hence $p = 0.09 + h$, where h is of the denomination thousandths. The coefficients 1, 6.27,... of the transformed equation for h appear in heavy type just under the first zigzag line in the following scheme:

$$
\begin{array}{lllll}
1 & 6 & 10 & -1 & \big|\,0.09 \\
  & 0.09 & 0.5481 & 0.949329 & \\
\hline
1 & 6.09 & 10.5481 & -0.050671 & \\
  & 0.09 & 0.5562 & & \\
\hline
1 & 6.18 & 11.1043 & & \quad 0.05 \\
  & 0.09 & & & \quad \overline{11.1} \\
\hline
1 & \mathbf{6.27} & & & = 0.004 \\
  & 0.004 & 0.025096 & 0.044517584 & \\
\hline
1 & 6.274 & 11.129396 & -0.006153416 & \\
  & 0.004 & 0.025112 & & \\
\hline
1 & 6.278 & 11.154508 & & \\
  & 0.004 & & & \\
\hline
1 & \mathbf{6.282} & & & \\
\end{array}
$$

Hence $x = 2.094 + t$, where t is a root of

$$t^3 + 6.282t^2 + 11.154508t - 0.006153416 = 0.$$

By the last two terms, t is between 0.0005 and 0.0006. Then the value of $C \equiv t^3 + 6.282t^2$ is found to lie between 0.00000157 and 0.00000227. Hence we may ignore C provided the constant term be reduced by an amount between these limits. Whichever of the two limits we use, we obtain the same dividend below correct to 6 decimal places.

$$
\begin{array}{c|c|c}
\overset{\times\ \times\times}{11.154508} & 0.006151 & 0.000551 = t \\
& 5577 & \\ \cline{2-2}
& 574 & \\
& 558 & \\ \cline{2-2}
& 16 & \\
& 11 & \\ \cline{2-2}
& 5 &
\end{array}
$$

Since the quotient is 0.0005+, only two decimal places of the divisor are used, except to see by inspection how much is to be carried when making the first multiplication. Hence we mark a cross above the figure 5 in the hundredths place of the divisor and use only 11.15. Before making the multiplication by the second significant figure 5 of the quotient t, we mark a cross over the figure 1 in the tenths place of the divisor and hence use only 11.1. Thus $x = 2.0945514+$, with doubt only as to whether the last figure should be 4 or 5.

If we require a greater number of decimal places, it is not necessary to go back and construct a new transformed equation from the equation in t. We have only to revise our preceding dividend on the basis of our present better value of t. We now know that t is between 0.000551 and 0.000552. To compute the new value of the correction C, in which we may evidently ignore t^3, we use logarithms.

$$
\begin{array}{llll}
\log & 5.51 = .74115 & \log & 5.52 = .74194 \\
\therefore \log & 5.51^2 = 1.48230 & \therefore \log & 5.52^2 = 1.48388 \\
\log & 6.282 = \underline{.79810} & \log & 6.282 = \underline{.79810} \\
\log & 190.72 = 2.28040 & \log & 191.42 = 2.28198
\end{array}
$$

Hence C is between 0.000001907 and 0.000001915. Whichever of the two limits we use, we obtain the same new dividend below correct to 8 decimal places.

$$\overset{\text{x xxxx}}{11.154508} \Big| \quad 0.00615150 \quad \Big| \; 0.00055148$$

$$557725$$

$$\overline{57425}$$
$$55773$$

$$\overline{1652}$$
$$1115$$

$$\overline{537}$$
$$446$$

$$\overline{91}$$
$$89$$

$$\overline{2}$$

Hence, finally, $x = 2.094551482$, with doubt only as to the last figure.

EXERCISES

(The number of transformations made by synthetic division should be about half the number of significant figures desired for a root.)

By one of the methods in Chapter VI, isolate each real root of the following equations, and compute each real root to 5 decimal places.

1. $x^3 + 2x + 20 = 0.$ 　　　　　**2.** $x^3 + 3x^2 - 2x - 5 = 0.$

3. $x^3 + x^2 - 2x - 1 = 0.$ 　　　　**4.** $x^4 + 4x^3 - 17.5x^2 - 18x + 58.5 = 0.$

5. $x^4 - 11,727x + 40,385 = 0.$ 　　**6.** $x^3 = 10.$

Find to 7 decimal places all the real roots of

7. $x^3 + 4x^2 - 7 = 0.$ 　　　　　**8.** $x^3 - 7x - 7 = 0.$

Find to 8 decimal places

9. The root between 2 and 3 of $x^3 - x - 9 = 0$ (make only 3 transformations).

10. The real cube root of 7.976.

11. The abscissa of the real point of intersection of the conics $y = x^2$, $xy + x + 3y - 6 = 0.$

12. Find to 3 decimal places the abscissas of the points of intersection of $x^2 + y^2 = 9$, $y = x^2 - x$.

13. A sphere two feet in diameter is formed of a kind of wood a cubic foot of which weighs two-thirds as much as a cubic foot of water (i.e., the specific gravity of the wood is $2/3$). Find to four significant figures the depth h to which the floating sphere will sink in water.

Hints: The volume of a sphere of radius r is $\frac{4}{3}\pi r^3$. Hence our sphere whose radius is 1 foot weighs as much as $\frac{4}{3}\pi \cdot \frac{2}{3}$ cubic feet of water. The volume of the submerged portion of the sphere is $\pi h^2(r - \frac{1}{3}h)$ cubic feet. Since this is also the volume of the displaced water, its value for $r = 1$ must equal $\frac{4}{3}\pi \cdot \frac{2}{3}$. Hence $h^3 - 3h^2 + \frac{8}{3} = 0$.

14. If the specific gravity of cork is $1/4$, find to four significant figures how far a cork sphere two feet in diameter will sink in water.

15. Compute $\cos 20°$ to four decimal places by use of

$$\cos 3A = 4\cos^3 A - 3\cos A, \qquad \cos 60° = \tfrac{1}{2}.$$

16. Three intersecting edges of a rectangular parallelopiped are of lengths 6, 8, and 10 feet. If the volume is increased by 300 cubic feet by equal elongations of the edges, find the elongation to three decimal places.

17. Given that the volume of a right circular cylinder is $\alpha\pi$ and the total area of its surface is $2\beta\pi$, prove that the radius r of its base is a root of $r^3 - \beta r + \alpha = 0$. If $\alpha = 56$, $\beta = 28$, find to four decimal places the two positive roots r. The corresponding altitude is α/r^2.

18. What rate of interest is implied in an offer to sell a house for $2700 cash, or in annual installments each of $1000 payable 1, 2, and 3 years from date?

Hint: The amount of $2700 with interest for 3 years should be equal to the sum of the first payment with interest for 2 years, the amount of the second payment with interest for 1 year, and the third payment. Hence if r is the rate of interest and we write x for $1 + r$, we have

$$2700x^3 = 1000x^2 + 1000x + 1000.$$

19. Find the rate of interest implied in an offer to sell a house for $3500 cash, or in annual installments each of $1000 payable 1, 2, 3, and 4 years from date.

20. Find the rate of interest implied in an offer to sell a house for $3500 cash, or $4000 payable in annual installments each of $1000, the first payable now.

75. Newton's Method. Prior to 1676, Newton[2] had already found the root between 2 and 3 of equation (1). He replaced x by $2+p$ and obtained (2). Since p is a decimal, he neglected the terms in p^3 and p^2, and hence obtained $p = 0.1$, approximately. Replacing p by $0.1 + q$ in (2), he obtained

$$q^3 + 6.3q^2 + 11.23q + 0.061 = 0.$$

Dividing -0.061 by 11.23, he obtained -0.0054 as the approximate value of q. Neglecting q^3 and replacing q by $-0.0054 + r$, he obtained

$$6.3r^2 + 11.16196r + 0.000541708 = 0.$$

Dropping $6.3r^2$, he found r and hence

$$x = 2 + 0.1 - 0.0054 - 0.00004853 = 2.09455147,$$

of which all figures but the last are correct (§74). But the method will not often lead so quickly to so accurate a value of the root.

Newton used the close approximation 0.1 to p, in spite of the fact that this value exceeds the root p and hence led to a negative correction at the next step. This is in contrast with Horner's method in which each correction is positive, so that each approximation must be chosen less than the root, as 0.09 for p.

Newton's method may be presented in the following general form, which is applicable to any equation $f(x) = 0$, whether $f(x)$ is a polynomial or not. Given an approximate value a of a real root, we can usually find a closer approximation $a + h$ to the root by neglecting the powers h^2, h^3,... of the small number h in Taylor's formula (§56)

$$f(a+h) = f(a) + f'(a)h + f''(a)\frac{h^2}{2} + \cdots$$

and hence by taking

$$f(a) + f'(a)h = 0, \qquad h = \frac{-f(a)}{f'(a)}.$$

We then repeat the process with $a_1 = a + h$ in place of the former a.

Thus in Newton's example, $f(x) = x^3 - 2x - 5$, we have, for $a = 2$,

$$h = \frac{-f(2)}{f'(2)} = \frac{1}{10}, \qquad a_1 = a + h = 2.1,$$

$$h_1 = \frac{-f(2.1)}{f'(2.1)} = \frac{-0.061}{11.23} = -0.0054\ldots.$$

[2]Isaac Newton, *Opuscula*, I, 1794, p. 10, p. 37.

76. Graphical Discussion of Newton's Method.   Using rectangular coördinates, consider the graph of $y = f(x)$ and the point P on it with the abscissa $OQ = a$ (Fig. 22). Let the tangent at P meet the x-axis at T and let

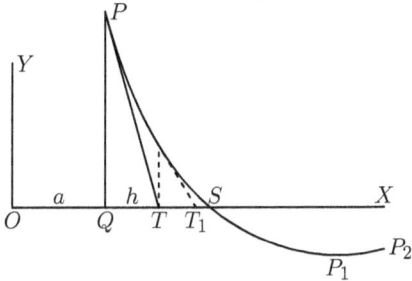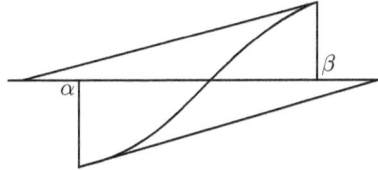

FIG. 22   FIG. 23

the graph meet the x-axis at S. Take $h = QT$, the subtangent. Then

$$QP = f(a), \qquad f'(a) = \tan XTP = \frac{-f(a)}{h},$$

$$h = \frac{-f(a)}{f'(a)}.$$

In the graph in Fig. 22, $OT = a + h$ is a better approximation to the root OS than $OQ = a$. The next step (indicated by dotted lines) gives a still better approximation OT_1.

If, however, we had begun with the abscissa a of a point P_1 in Fig. 22 near a bend point, the subtangent would be very large and the method would probably fail to give a better approximation. Failure is certain if we use a point P_2 such that a single bend point lies between it and S.

We are concerned with the approximation to a root previously isolated as the only real root between two given numbers α and β. These should be chosen so nearly equal that $f'(x) = 0$ has no real root between α and β, and hence $f(x) = y$ has no bend point between α and β. Further, if $f''(x) = 0$ has a root between our limits, our graph will have an inflexion point with an abscissa between α and β, and the method will likely fail (Fig. 23).

Let, therefore, neither $f'(x)$ nor $f''(x)$ vanish between α and β. Since f'' preserves its sign in the interval from α to β, while f changes in sign, f'' and f will have the same sign for one end point. According as the abscissa of this point is α or β, we take $a = \alpha$ or $a = \beta$ for the first step of Newton's process. In fact, the tangent at one of the end points meets the x-axis at a point T with an abscissa within the interval from α to β. If $f'(x)$ is positive in the interval, so that the tangent makes an acute angle with the x-axis, we have Fig. 24 or Fig. 25; if f' is negative, Fig. 26 or Fig. 22.

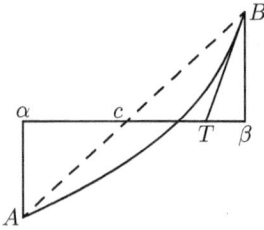

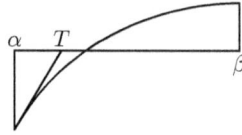

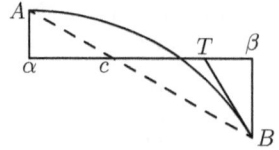

FIG. 24   FIG. 25   FIG. 26

In Newton's example, the graph between the points with the abscissas $\alpha = 2$ and $\beta = 3$ is of the type in Fig. 24, but more nearly like a vertical straight line. In view of this feature of the graph, we may safely take $a = \alpha$, as did Newton, although our general procedure would be to take $a = \beta$. The next step, however, accords with our present process; we have $\alpha = 2$, $\beta = 2.1$ in Fig. 24 and hence we now take $a = \beta$, getting

$$\frac{0.061}{11.23} = 0.0054$$

as the subtangent, and hence $2.1 - 0.0054$ as the approximate root.

If we have secured (as in Fig. 24 or Fig. 26) a better upper limit to the root than β, we may take the abscissa c of the intersection of the chord AB with the x-axis as a better lower limit than α. By similar triangles,

$$-f(\alpha) : c - \alpha = f(\beta) : \beta - c,$$

whence

(3)
$$c = \frac{\alpha f(\beta) - \beta f(\alpha)}{f(\beta) - f(\alpha)}.$$

This method of finding the value of c intermediate to α and β is called the method of interpolation (*regula falsi*).

In Newton's example, $\alpha = 2$, $\beta = 2.1$,

$$f(\alpha) = -1, \qquad f(\beta) = 0.061, \qquad c = 2.0942.$$

The advantage of having c at each step is that we know a close limit of the error made in the approximation to the root.

We may combine the various possible cases discussed into one:

If $f(x) = 0$ has a single real root between α and β, and $f'(x) = 0$, $f''(x) = 0$ have no real root between α and β, and if we designate by β that one of the numbers α and β for which $f(\beta)$ and $f''(\beta)$ have the same sign, then the root lies in the narrower interval from c to $\beta - f(\beta)/f'(\beta)$, where c is given by (3).

It is possible to prove[3] this theorem algebraically and to show that by repeated applications of it we can obtain two limits α', β' between which the root lies, such that $\alpha' - \beta'$ is numerically less than any assigned positive number. Hence the root can be found in this manner to any desired accuracy.

EXAMPLE. $f(x) = x^3 - 2x^2 - 2$, $\alpha = 2\frac{1}{4}$, $\beta = 2\frac{1}{2}$. Then

$$f(\alpha) = -\tfrac{47}{64}, \qquad f(\beta) = \tfrac{9}{8}.$$

Neither of the roots 0, 4/3 of $f'(x) = 0$ lies between α and β, so that $f(x) = 0$ has a single real root between these limits (§65). Nor is the root $\frac{2}{3}$ of $f''(x) = 0$ within these limits. The conditions of the theorem are therefore satisfied. For $\alpha < x < \beta$, the graph is of the type in Fig. 24. We find that approximately

$$c = \tfrac{559}{238} = 2.3487, \qquad \beta_1 = \beta - \frac{f(\beta)}{f'(\beta)} = 2.3714,$$

$$\beta_1 - \frac{f(\beta_1)}{f'(\beta_1)} = 2.3597.$$

For $x = 2.3593$, $f(x) = -0.00003$. We therefore have the root to four decimal places. For $a = 2.3593$,

$$f'(a) = 7.2620, \qquad a - \frac{f(a)}{f'(a)} = 2.3593041,$$

which is the value of the root correct to 7 decimal places. We at once verify that the result is greater than the root in view of our work and Fig. 24, while if we change the final digit from 1 to 0, $f(x)$ is negative.

EXERCISES

1. For $f(x) = x^4 + x^3 - 3x^2 - x - 4$, show by Descartes' rule of signs that $f'(x) = 0$ and $f''(x) = 0$ each have a single positive root and that neither has a root between 1 and 2. Which of the values 1 and 2 should be taken as β?

2. When seeking a root between 2 and 3 of $x^3 - x - 9 = 0$, which value should be taken as β?

---

[3]Weber's *Algebra*, 2d ed., I, pp. 380–382; *Kleines Lehrbuch der Algebra*, 1912, p. 163.

77. Systematic Computation of Roots by Newton's Method.   By way of illustration we shall compute to 7 decimal places a positive root of

$$f(x) = x^4 + x^3 - 3x^2 - x - 4 = 0.$$

Since $f(1) = -6$, $f(2) = 6$, there is a real root between 1 and 2. Since

$$f'(x) = 4x^3 + 3x^2 - 6x - 1, \qquad f'(1) = 0, \qquad f'(2) = 31,$$

the graph of $y = f(x)$ is approximately horizontal near $(1, -6)$ and approximately vertical near $(2, 6)$. Hence the root is much nearer to 2 than to 1. Thus in applying Newton's method we employ $a = 2$ as the first approximation to the root. The correction h is then

$$h = \frac{-f(2)}{f'(2)} = \frac{-6}{31} = -0.2 \ldots.$$

The work of performing the substitutions $x = 2 + d$, $d = -0.2 + e, \ldots$, to find the transformed equations satisfied by d, $e, \ldots$, is done by synthetic division, exactly as in Horner's method, except that some of the multipliers are now negative: see Table 1, page 107.

The root is $2 - 0.2 - 0.04 + 0.000302 = 1.760302$, in which the last figure is in slight doubt. Indeed, it can be proved that *if the final fraction g, when expressed as a decimal, has k zeros between the decimal point and the first significant figure, the division may be safely carried to 2k decimal places.* In our example $k = 3$, so that we retained 6 decimal places in g.

To proceed independently of this rule, we note that g is obviously between 0.00030 and 0.00031. Then the value of $g^4 + 8.04g^3 + 20.8656g^2$ is found to lie between 0.000001878 and 0.000002006. Whichever of these limits we use as a correction by which to reduce the constant term, we obtain the same dividend below correct to 6 decimal places.

$$
\begin{array}{c|c|c}
\overset{\times\times\ \times\times}{19.539904} & 0.005896 & 0.0003017 \\
& 005862 & \\
\hline
& 34 & \\
& 20 & \\
\hline
& 14 & \\
& 14 & \\
\hline
\end{array}
$$

Hence the root is 1.7603017 to 7 decimal places.

1	1	−3	−1	−4	$\underline{2}$
	2	6	6	10	
1	3	3	5	**6**	
	2	10	26		
1	5	13	**31**		
	2	14			
1	7	**27**			
	2				
1	9				$\underline{-0.2}$
	−0.2	−1.76	−5.048	−5.1904	
	8.8	25.24	25.952	**0.8096**	
	−0.2	−1.72	−4.704		
	8.6	23.52	**21.248**		
	−0.2	−1.68			
	8.4	**21.84**			
	−0.2				
1	**8.2**				
	−0.04	−0.3264	−0.860544	−0.81549824	
	8.16	21.5136	20.387456	**−0.00589824**	
	−0.04	−0.3248	−0.847552		
	8.12	21.1888	**19.539904**		
	−0.04	−0.3232			
	8.08	**20.8656**			
	−0.04				
1	**8.04**				

$$\frac{-0.8096}{21.248} = -0.04$$

$$g = \frac{0.005898}{19.54} = .000302$$

TABLE 1.

EXERCISES

1. Find to 8 decimal places the root between 2 and 3 of $x^3 - x - 9 = 0$.

2. Find to 7 decimal places the root between 2 and 3 of $x^3 - 2x^2 - 2 = 0$.

3. Find the real cube root of 7.976 to 6 decimal places.

4. Explain by Taylor's expansion of $f(2 + d)$ why the values of

$$f(2), \qquad f'(2), \qquad \tfrac{1}{2}f''(2), \qquad \frac{1}{2 \cdot 3}f'''(2), \qquad \frac{1}{2 \cdot 3 \cdot 4}f''''(2)$$

are in reverse order the coefficients of the transformed equation

$$d^4 + 9d^3 + 27d^2 + 31d + 6 = 0,$$

obtained in the Example in the text, and printed in heavy type.

5. The method commonly used to find the positive square root of n by a computing machine consists in dividing n by an assumed approximate value a of the square root and taking half the sum of a and the quotient as a better approximation. Show that the latter agrees with the value of $a + h$ given by applying Newton's method to $f(x) = x^2 - n$.

78. Newton's Method for Functions not Polynomials.

EXAMPLE 1.   Find the angle x at the center of a circle subtended by a chord which cuts off a segment whose area is one-eighth of that of the circle.

Solution.   If x is measured in radians and if r is the radius, the area of the segment is equal to the left member of

$$\tfrac{1}{2}r^2(x - \sin x) = \tfrac{1}{8}\pi r^2,$$

whence

$$x - \sin x = \tfrac{1}{4}\pi.$$

By means of a graph of $y = \sin x$ and the straight line represented by $y = x - \tfrac{1}{4}\pi$, we see that the abscissa of their point of intersection is approximately 1.78 radians or 102°. Thus $a = 102°$ is a first approximation to the root of

$$f(x) \equiv x - \sin x - \tfrac{1}{4}\pi = 0.$$

By Newton's method a better approximation is $a + h$, where[4]

$$h = \frac{-f(a)}{f'(a)} = \frac{-a + \sin a + \tfrac{1}{4}\pi}{1 - \cos a}.$$

---

[4] The derivative of $\sin x$ is $\cos x$. We need the limit of

$$\frac{\sin(x + 2k) - \sin x}{2k} \equiv \frac{2\cos\tfrac{1}{2}(2x + 2k)\sin\tfrac{1}{2}(2k)}{2k} \equiv \frac{\cos(x + k)\sin k}{k}$$

as $2k$ approaches zero. Since the ratio of $\sin k$ to k approaches 1, the limit is $\cos x$.

$$\begin{aligned} \sin 102° &= \quad 0.9781 \\ \tfrac{1}{4}(3.1416) &= \quad 0.7854 \\ &\qquad \overline{1.7635} \end{aligned} \qquad\qquad \begin{aligned} \cos 102° &= -0.2079 \\ 1 - \cos 102° &= \quad 1.2079 \end{aligned}$$

$$102° = 1.7802 \text{ radians} \qquad\qquad h = \frac{-0.0167}{1.2079} = -0.0138$$
$$\underline{\quad -0.0167\quad} \qquad\qquad a_1 = a + h = 1.7664$$

$$h_1 = \frac{-f(a_1)}{f'(a_1)} = \frac{-1.7664 + 0.9809 + 0.7854}{1.1944} = -0.0001.$$

Hence $x = a_1 + h_1 = 1.7663$ radians, or $101°12'$.

EXAMPLE 2.[5] Solve $x - \log x = 7$, the logarithm being to base 10.

Solution. Evidently x exceeds 7 by a positive decimal which is the value of $\log x$. Hence in a table of common logarithms, we seek a number x between 7 and 8 whose logarithm coincides approximately with the decimal part of x. We read off the values in the second column.

x	$\log x$	$x - \log x$
7.897	0.89746	6.99954
7.898	0.89752	7.00048

By the final column the ratio of interpolation is 46/94. Hence $x = 7.8975$ to four decimal places.

EXAMPLE 3. Solve $2x - \log x = 7$, the logarithm being to base 10.

Solution. Evidently x is a little less than 4. A table of common logarithms shows at once that a fair approximation to x is $a = 3.8$. Write

$$f(x) \equiv 2x - \log x - 7, \qquad \log x = M \log_e x, \qquad M = 0.4343.$$

By calculus, the derivative of $\log_e x$ is $1/x$. Hence

$$f'(x) = 2 - \frac{M}{x}, \qquad f'(a) = 2 - 0.1143 = 1.8857,$$
$$f(a) = 0.6 - \log 3.8 = 0.6 - 0.57978 = 0.02022,$$
$$-h = \frac{f(a)}{f'(a)} = 0.0107, \qquad a_1 = a + h = 3.7893,$$
$$f(a_1) = 0.000041, \qquad f(3.7892) = -0.000148.$$
$$\frac{148}{189} \times 0.0001 = 0.000078, \qquad x = 3.789278.$$

All figures of x are correct as shown by Vega's table of logarithms to 10 places.

---

[5]This Ex. 2, which should be contrasted with Ex. 3, is solved by interpolation since that method is simpler than Newton's method in this special case.

EXERCISES

Find the angle x at the center of a circle subtended by a chord which cuts off a segment whose ratio to the circle is

1. $\frac{1}{4}$.   **2.** $\frac{3}{8}$.

When the logarithms are to base 10,

3. Solve $2x - \log x = 9$.   **4.** Solve $3x - \log x = 9$.

5. Find the angle just $> 15°$ for which $\frac{1}{2}\sin x + \sin 2x = 0.64$.

6. Find the angle just $> 72°$ for which $x - \frac{1}{2}\sin x = \frac{1}{4}\pi$.

7. Find all solutions of Ex. 5 by replacing $\sin 2x$ by $2\sin x \cos x$, squaring, and solving the quartic equation for $\cos x$.

8. Solve similarly $\sin x + \sin 2x = 1.2$.

9. Find x to 6 decimal places in $\sin x = x - 2$.

10. Find x to 5 decimal places in $x = 3\log_e x$.

79. Imaginary Roots. To find the imaginary roots $x + yi$ of an equation $f(z) = 0$ with real coefficients, expand $f(x + yi)$ by Taylor's theorem; we get

$$f(x) + f'(x)yi - f''(x)\frac{y^2}{1\cdot 2} - f'''(x)\frac{y^3 i}{1\cdot 2\cdot 3} + \cdots = 0.$$

Since x and y are to be real, and $y \neq 0$,

(4)
$$\begin{cases} f(x) - f''(x)\dfrac{y^2}{1\cdot 2} + f''''(x)\dfrac{y^4}{1\cdot 2\cdot 3\cdot 4} - \cdots = 0, \\ f'(x) - f'''(x)\dfrac{y^2}{1\cdot 2\cdot 3} + f^{(5)}(x)\dfrac{y^4}{5!} - \cdots = 0. \end{cases}$$

In the Example and Exercises below, $f(z)$ is of degree 4 or less. Then the second equation (4) is linear in y^2. Substituting the resulting value of y^2 in the first equation (4), we obtain an equation $E(x) = 0$, whose real roots may be found by one of the preceding methods. If the degree of $f(z)$ exceeds 4, we may find $E(x) = 0$ by eliminating y^2 between the two equations (4) by one of the methods to be explained in Chapter X.

EXAMPLE.   For $f(z) = z^4 - z + 1$, equations (4) are

$$x^4 - x + 1 - 6x^2 y^2 + y^4 = 0, \qquad 4x^3 - 1 - 4xy^2 = 0.$$

Thus

$$y^2 = x^2 - \frac{1}{4x}, \qquad -4x^6 + x^2 + \frac{1}{16} = 0.$$

The cubic equation in x^2 has the single real root

$$x^2 = 0.528727, \qquad x = \pm 0.72714.$$

Then $y^2 = 0.184912$ or 0.87254, and

$$z = x + yi = 0.72714 \pm 0.43001i, \qquad -0.72714 \pm 0.93409i.$$

EXERCISES

Find the imaginary roots of

1. $z^3 - 2z - 5 = 0.$

2. $28z^3 + 9z^2 - 1 = 0.$

3. $z^4 - 3z^2 - 6z = 2.$

4. $z^4 - 4z^3 + 11z^2 - 14z + 10 = 0.$

5. $z^4 - 4z^3 + 9z^2 - 16z + 20 = 0.$ Hint:

$$E(x) \equiv x(x - 2)(16x^4 - 64x^3 + 136x^2 - 144x + 65) = 0,$$

and the last factor becomes $(w^2 + 1)(w^2 + 9)$ for $2x = w + 2$.

NOTE.   If we know a real root r of a cubic equation $f(z) = 0$, we may remove the factor $z - r$ and solve the resulting quadratic equation. When, as usual, r involves several decimal places, this method is laborious and unsatisfactory. But we may utilize a device, explained in the author's *Elementary Theory of Equations*, pp. 119–121, §§6, 7. As there explained, a similar device may be used when we know two real roots of a quartic equation.

MISCELLANEOUS EXERCISES

(Give answers to 6 decimal places, unless the contrary is stated.)

1. What arc of a circle is double its chord?

2. What arc of a circle is double the distance from the center of the circle to the chord of the arc?

3. If A and B are the points of contact of two tangents to a circle of radius unity from a point P without it, and if arc AB is equal to PA, find the length of the arc.

4. Find the angle at the center of a circle of a sector which is bisected by its chord.

5. Find the radius of the smallest hollow iron sphere, with air exhausted, which will float in water if its shell is 1 inch thick and the specific gravity of iron is 7.5.

6. From one end of a diameter of a circle draw a chord which bisects the semi-circle.

7. The equation $x \tan x = c$ occurs in the theory of vibrating strings. Its approximate solutions may be found from the graphs $y = \cot x$, $y = x/c$. Find x when $c = 1$.

8. The equation $\tan x = x$ occurs in the study of the vibrations of air in a spherical cavity. From an approximate solution $x_1 = 1.5\pi$, we obtain successively better approximations $x_2 = \tan^{-1} x_1 = 1.4334\pi$, $x_3 = \tan^{-1} x_2, \ldots$. Find the first three solutions to 4 decimal places.

9. Find to 3 decimal places the first five solutions of

$$\tan x = \frac{2x}{2 - x^2},$$

which occurs in the theory of vibrations in a conical pipe.

10. $4\tau x^3 - (3x - 1)^2 = 0$ arises in the study of the isothermals of a gas. Find its roots when (i) $\tau = 0.002$ and (ii) $\tau = 0.99$.

11. Solve $x^x = 100$.

12. Solve $x = 10 \log x$.

13. Solve $x + \log x = x \log x$.

14. Solve Kepler's equation $M = x - e \sin x$ when $M = 332°28'54.8''$, $e = 14°3'20''$.

15. In what time would a sum of money at 6% interest compounded annually amount to as much as the same sum at simple interest at 8%?

16. In a semicircle of diameter x is inscribed a quadrilateral with sides a, b, c, x; then $x^3 - (a^2 + b^2 + c^2)x - 2abc = 0$ (I. Newton). Given $a = 2$, $b = 3$, $c = 4$, find x.

17. What rate of interest is implied in an offer to sell a house for $9000 cash, or $1000 down and $3000 at the end of each year for three years?

CHAPTER VIII

80. Solution of Two Linear Equations by Determinants of Order 2.
Assume that there is a pair of numbers x and y for which

$$(1) \qquad \begin{cases} a_1x + b_1y = k_1, \\ a_2x + b_2y = k_2. \end{cases}$$

Multiply the members of the first equation by b_2 and those of the second equation by $-b_1$, and add the resulting equations. We get

$$(a_1b_2 - a_2b_1)x = k_1b_2 - k_2b_1.$$

Employing the respective multipliers $-a_2$ and a_1, we get

$$(a_1b_2 - a_2b_1)y = a_1k_2 - a_2k_1.$$

The common multiplier of x and y is

$$(2) \qquad a_1b_2 - a_2b_1,$$

and is denoted by the symbol

$$(2') \qquad \begin{vmatrix} a_1 & b_1 \\ a_2 & b_2 \end{vmatrix},$$

which is called a *determinant of the second order*, and also called the determinant of the coefficients of x and y in equations (1). The results above may now be written in the form

$$(3) \qquad \begin{vmatrix} a_1 & b_1 \\ a_2 & b_2 \end{vmatrix} x = \begin{vmatrix} k_1 & b_1 \\ k_2 & b_2 \end{vmatrix}, \qquad \begin{vmatrix} a_1 & b_1 \\ a_2 & b_2 \end{vmatrix} y = \begin{vmatrix} a_1 & k_1 \\ a_2 & k_2 \end{vmatrix}.$$

We shall call k_1 and k_2 the known terms of our equations (1). Hence, *if D is the determinant of the coefficients of the unknowns, the product of D by any one* *of the unknowns is equal to the determinant obtained from D by substituting the known terms in place of the coefficients of that unknown.*

If $D \neq 0$, relations (3) uniquely determine values of x and y:

$$x = \frac{k_1 b_2 - k_2 b_1}{D}, \qquad y = \frac{a_1 k_2 - a_2 k_1}{D},$$

and these values satisfy equations (1); for example,

$$a_1 x + b_1 y = \frac{(a_1 b_2 - a_2 b_1) k_1}{D} = k_1.$$

Hence our equations (1) have been solved by determinants when $D \neq 0$. We shall treat in §96 the more troublesome case in which $D = 0$.

EXAMPLE.   For $2x - 3y = -4$, $6x - 2y = 2$, we have

$$\begin{vmatrix} 2 & -3 \\ 6 & -2 \end{vmatrix} x = \begin{vmatrix} -4 & -3 \\ 2 & -2 \end{vmatrix}, \qquad 14x = 14, \qquad x = 1,$$

$$14y = \begin{vmatrix} 2 & -4 \\ 6 & 2 \end{vmatrix} = 28, \qquad y = 2.$$

EXERCISES

Solve by determinants the following systems of equations:

1. $8x - y = 34,$
    $x + 8y = 53.$

2. $3x + 4y = 10,$
    $4x + y = 9.$

3. $ax + by = a^2,$
    $bx - ay = ab.$

81. Solution of Three Linear Equations by Determinants of Order 3.

Consider a system of three linear equations

$$(4) \qquad \begin{aligned} a_1 x + b_1 y + c_1 z &= k_1, \\ a_2 x + b_2 y + c_2 z &= k_2, \\ a_3 x + b_3 y + c_3 z &= k_3. \end{aligned}$$

Multiply the members of the first, second and third equations by

$$(5) \qquad b_2 c_3 - b_3 c_2, \qquad b_3 c_1 - b_1 c_3, \qquad b_1 c_2 - b_2 c_1,$$

respectively, and add the resulting equations. We obtain an equation in which the coefficients of y and z are found to be zero, while the coefficient of x is

$$(6) \qquad a_1 b_2 c_3 - a_1 b_3 c_2 + a_2 b_3 c_1 - a_2 b_1 c_3 + a_3 b_1 c_2 - a_3 b_2 c_1.$$

Such an expression is called a *determinant of the third order* and denoted by the symbol

(6′)
$$\begin{vmatrix} a_1 & b_1 & c_1 \\ a_2 & b_2 & c_2 \\ a_3 & b_3 & c_3 \end{vmatrix}.$$

The nine numbers $a_1, \ldots, c_3$ are called the *elements* of the determinant. In the symbol these elements lie in three (horizontal) *rows*, and also in three (vertical) *columns*. Thus a_2, b_2, c_2 are the elements of the second row, while the three c's are the elements of the third column.

The equation (free of y and z), obtained above, may now be written as

$$\begin{vmatrix} a_1 & b_1 & c_1 \\ a_2 & b_2 & c_2 \\ a_3 & b_3 & c_3 \end{vmatrix} x = \begin{vmatrix} k_1 & b_1 & c_1 \\ k_2 & b_2 & c_2 \\ k_3 & b_3 & c_3 \end{vmatrix},$$

since the right member was the sum of the products of the expressions (5) by k_1, k_2, k_3, and hence may be derived from (6) by replacing the a's by the k's. Thus the theorem of §80 holds here as regards the unknown x. We shall later prove, without the laborious computations just employed, that the theorem holds for all three unknowns.

82. The Signs of the Terms of a Determinant of Order 3.   In the six terms of our determinant (6), the letters a, b, c were always written in this sequence, while the subscripts are the six possible arrangements of the numbers 1, 2, 3. The first term $a_1 b_2 c_3$ shall be called the *diagonal term*, since it is the product of the elements in the main diagonal running from the upper left-hand corner to the lower right-hand corner of the symbol (6′) for the determinant. The subscripts in the term $-a_1 b_3 c_2$ are derived from those of the diagonal term by interchanging 2 and 3, and the minus sign is to be associated with the fact that an odd number (here one) of interchanges of subscripts were used. To obtain the arrangement 2, 3, 1 of the subscripts in the term $+a_2 b_3 c_1$ from the natural order 1, 2, 3 (in the diagonal term), we may first interchange 1 and 2, obtaining the arrangement 2, 1, 3, and then interchange 1 and 3; an even number (two) of interchanges of subscripts were used and the sign of the term is plus.

While the arrangement 1, 3, 2 was obtained from 1, 2, 3 by one interchange $(2, 3)$, we may obtain it by applying in succession the three interchanges $(1, 2)$, $(1, 3)$, $(1, 2)$, and in many new ways. To show that the number of interchanges which will produce the final arrangement 1, 3, 2 is odd in every case, note that each of the three possible interchanges, viz., $(1,2)$, $(1,3)$, and $(2,3)$, changes the sign of the product

$$P = (x_1 - x_2)(x_1 - x_3)(x_2 - x_3),$$

where the x's are arbitrary variables. Thus a succession of k interchanges yields P or $-P$ according as k is even or odd. Starting with the arrangement 1, 2, 3 and applying k successive interchanges, suppose that we obtain the final arrangement 1, 3, 2. But if in P we replace the subscripts 1, 2, 3 by 1, 3, 2, respectively, i.e., if we interchange 2 and 3, we obtain $-P$. Hence k is odd. We have therefore proved the following rule of signs:

Although the arrangement r, s, t of the subscripts in any term $\pm a_r b_s c_t$ of the determinant may be obtained from the arrangement 1, 2, 3 by various successions of interchanges, the number of these interchanges is either always an even number and then the sign of the term is plus or always an odd number and then the sign of the term is minus.

EXERCISES

Apply the rule of signs to all terms of

1. Determinant (6).   **2.** Determinant $a_1 b_2 - a_2 b_1$.

83. Number of Interchanges always Even or always Odd.   We now extend the result in §82 to the case of n variables $x_1, \ldots, x_n$. The product of all of their differences $x_i - x_j$ $(i < j)$ is

$$P = (x_1 - x_2)(x_1 - x_3) \cdots (x_1 - x_n)$$
$$\cdot (x_2 - x_3) \cdots (x_2 - x_n)$$
$$\vdots$$
$$\cdot (x_{n-1} - x_n).$$

Interchange any two subscripts i and j. The factors which involve neither i nor j are unaltered. The factor $(x_i - x_j)$ involving both is changed in sign. The remaining factors may be paired to form the products

$$\pm (x_i - x_k)(x_j - x_k)   (k = 1, \ldots, n;   k \neq i, k \neq j).$$

Such a product is unaltered. Hence P is changed in sign.

Suppose that an arrangement $i_1, i_2, \ldots, i_n$ can be obtained from $1, 2, \ldots, n$ by using m successive interchanges and also by t successive interchanges.

Make these interchanges on the subscripts in P; the resulting functions are equal to $(-1)^m P$ and $(-1)^t P$, respectively. But the resulting functions are identical since either can be obtained at one step from P by replacing the subscript 1 by i_1, 2 by i_2, ..., n by i_n. Hence

$$(-1)^m P \equiv (-1)^t P,$$

so that m and t are both even or both odd.

Thus *if the same arrangement is derived from* 1, 2, ..., n *by* m *successive interchanges as by* t *successive interchanges, then* m *and* t *are both even or both odd.*

84. Definition of a Determinant of Order n.   We define a determinant of order 4 to be

(7)
$$\begin{vmatrix} a_1 & b_1 & c_1 & d_1 \\ a_2 & b_2 & c_2 & d_2 \\ a_3 & b_3 & c_3 & d_3 \\ a_4 & b_4 & c_4 & d_4 \end{vmatrix} = \sum_{(24)} \pm a_q b_r c_s d_t,$$

where q, r, s, t is any one of the 24 arrangements of $1, 2, 3, 4$, and the sign of the corresponding term is $+$ or $-$ according as an even or odd number of interchanges are needed to derive this arrangement q, r, s, t from $1, 2, 3, 4$. Although different numbers of interchanges will produce the same arrangement q, r, s, t from $1, 2, 3, 4$, these numbers are all even or all odd, as just proved, so that the sign is fully determined.

We have seen that the analogous definitions of determinants of orders 2 and 3 lead to our earlier expressions (2) and (6).

We will have no difficulty in extending the definition to a determinant of general order n as soon as we decide upon a proper notation for the n^2 elements. The subscripts $1, 2, \ldots, n$ may be used as before to specify the rows. But the alphabet does not contain n letters with which to specify the columns. The use of $e', e'', \ldots, e^{(n)}$ for this purpose would conflict with the notation for derivatives and besides be very awkward when exponents are used. It is customary in mathematical journals and scientific books (a custom not always followed in introductory text books, to the distinct disadvantage of the reader) to denote the n letters used to distinguish the n columns by $e_1, e_2, \ldots, e_n$ (or some other letter with the same subscripts) and to prefix (but see §85) such a subscript by the new subscript indicating the row. The symbol for the determinant is therefore

(8)
$$D = \begin{vmatrix} e_{11} & e_{12} & \cdots & e_{1n} \\ e_{21} & e_{22} & \cdots & e_{2n} \\ \cdots\cdots\cdots\cdots\cdots \\ e_{n1} & e_{n2} & \cdots & e_{nn} \end{vmatrix}.$$

By definition this shall mean the sum of the $n(n-1)\cdots 2\cdot 1$ terms

(9)
$$(-1)^i e_{i_1 1} e_{i_2 2} \cdots e_{i_n n}$$

in which $i_1, i_2, \ldots, i_n$ is an arrangement of $1, 2, \ldots, n$, derived from $1, 2, \ldots, n$ by i interchanges. Any term (9) of the determinant (8) is, apart from sign, the product of n factors, one and only one from each column, and one and only one from each row.

For example, if we take $n = 4$ and write a_j, b_j, c_j, d_j for $e_{j1}, e_{j2}, e_{j3}, e_{j4}$, the symbol (8) becomes (7) and the general term (9) becomes the general term $(-1)^i a_{i_1} b_{i_2} c_{i_3} d_{i_4}$ of the second member of (7).

EXERCISES

1. Find the six terms involving a_2 in the determinant (7).

2. What are the signs of $a_3 b_5 c_2 d_1 e_4$, $a_5 b_4 c_3 d_2 e_1$ in a determinant of order five?

3. Show that the arrangement $4, 1, 3, 2$ may be obtained from $1, 2, 3, 4$ by use of the two successive interchanges $(1, 4)$, $(1, 2)$, and also by use of the four successive interchanges $(1, 4)$, $(1, 3)$, $(1, 2)$, $(2, 3)$.

4. Write out the six terms of (8) for $n = 3$, rearrange the factors of each term so that the new first subscripts shall be in the order $1, 2, 3$, and verify that the resulting six terms are those of the determinant D' in §85 for $n = 3$.

85. Interchange of Rows and Columns. *Any determinant is not altered in value if in its symbol we replace the elements of the first, second, $\ldots, n$th rows by the elements which formerly appeared in the same order in the first, second, $\ldots, n$th columns, or briefly if we interchange the corresponding rows and columns. For example,*

$$\begin{vmatrix} a & b \\ c & d \end{vmatrix} = ad - bc = \begin{vmatrix} a & c \\ b & d \end{vmatrix}.$$

We are to prove that the determinant D given by (8) is equal to

$$D' = \begin{vmatrix} e_{11} & e_{21} & \cdots & e_{n1} \\ e_{12} & e_{22} & \cdots & e_{n2} \\ \cdots\cdots\cdots\cdots\cdots \\ e_{1n} & e_{2n} & \cdots & e_{nn} \end{vmatrix}.$$

If we give to D' a more familiar aspect by writing $e_{ik} = a_{ki}$ for each element so that, as in (8), the row subscript precedes instead of follows the column subscript, the definition of the determinant in terms of the a's gives D' in terms of the e's as the sum of all expressions

$$(-1)^i e_{1k_1} e_{2k_2} \cdots e_{nk_n},$$

in which $k_1, k_2, \ldots, k_n$ is an arrangement of $1, 2, \ldots, n$, derived from the latter sequence by i interchanges.

As for the terms of D, without altering (9), we may rearrange its factors so that the first subscripts shall appear in the order $1, 2, \ldots, n$, and obtain

$$(-1)^i e_{1k_1} e_{2k_2} \cdots e_{nk_n}.$$

This can be done by performing in reverse order the i successive interchanges of the letters e corresponding to the i successive interchanges which were used to derive the arrangement $i_1, i_2, \ldots, i_n$ of the first subscripts from the arrangement $1, 2, \ldots, n$. Thus the new second subscripts $k_1, \ldots, k_n$ are derived from the old second subscripts $1, \ldots, n$ by i interchanges. The resulting signed product is therefore a term of D'. Hence $D = D'$.

86. Interchange of Two Columns.   *A determinant is merely changed in sign by the interchange of any two of its columns.* For example,

$$D = \begin{vmatrix} a & b \\ c & d \end{vmatrix} = ad - bc, \qquad \Delta = \begin{vmatrix} b & a \\ d & c \end{vmatrix} = bc - ad = -D.$$

Let Δ be the determinant derived from (8) by the interchange of the rth and sth columns. The terms of Δ are therefore obtained from the terms (9) of D by interchanging r and s in the series of second subscripts. Interchange the rth and sth letters e to restore the second subscripts to their natural order. Since the first subscripts have undergone an interchange, the negative of any term of Δ is a term of D, and $\Delta = -D$.

87. Interchange of Two Rows.   *A determinant D is merely changed in sign by the interchange of any two rows.*

Let Δ be the determinant obtained from D by interchanging the rth and sth rows. By interchanging the rows and columns in D and in Δ, we get two determinants D' and Δ', either of which may be derived from the other by the interchange of the rth and sth columns. Hence, by §§85, 86,

$$\Delta = \Delta' = -D' = -D.$$

88. Two Rows or Two Columns Alike.   *A determinant is zero if any two of its rows or any two of its columns are alike.*

For, by the interchange of the two like rows or two like columns, the determinant is evidently unaltered, and yet must change in sign by §§86, 87. Hence $D = -D, D = 0.$

EXERCISES

1. Prove that the equation of the straight line determined by the two distinct points (x_1, y_1) and (x_2, y_2) is

$$\begin{vmatrix} x & y & 1 \\ x_1 & y_1 & 1 \\ x_2 & y_2 & 1 \end{vmatrix} = 0.$$

2. Show that

$$\begin{vmatrix} a_1 & b_1 & c_1 \\ a_2 & b_2 & c_2 \\ a_3 & b_3 & c_3 \end{vmatrix} = \begin{vmatrix} a_2 & c_2 & b_2 \\ a_1 & c_1 & b_1 \\ a_3 & c_3 & b_3 \end{vmatrix} = \begin{vmatrix} a_3 & a_1 & a_2 \\ b_3 & b_1 & b_2 \\ c_3 & c_1 & c_2 \end{vmatrix}.$$

By use of the Factor Theorem (§14) and the diagonal term, prove that

3.

$$\begin{vmatrix} 1 & 1 & 1 \\ a & b & c \\ a^2 & b^2 & c^2 \end{vmatrix} = (b - a)(c - a)(c - b).$$

4.

$$\begin{vmatrix} 1 & 1 & \cdots & 1 \\ x_1 & x_2 & \cdots & x_n \\ x_1^2 & x_2^2 & \cdots & x_n^2 \\ \cdots & \cdots & \cdots & \cdots \\ x_1^{n-1} & x_2^{n-1} & \cdots & x_n^{n-1} \end{vmatrix} = \prod_{\substack{i,j=1 \\ i>j}}^{n} (x_i - x_j).$$

This is known as the determinant of Vandermonde, who discussed it in 1770. The symbol on the right means the product of all factors of the type indicated.

5. Prove that a skew-symmetric determinant of odd order is zero:

$$\begin{vmatrix} 0 & a & b \\ -a & 0 & c \\ -b & -c & 0 \end{vmatrix} = 0, \qquad \begin{vmatrix} 0 & a & b & c & d \\ -a & 0 & e & f & g \\ -b & -e & 0 & h & j \\ -c & -f & -h & 0 & k \\ -d & -g & -j & -k & 0 \end{vmatrix} = 0.$$

89. Minors. The determinant of order $n - 1$ obtained by erasing (or covering up) the row and column crossing at a given element of a determinant of order n is called the *minor* of that element.

For example, in the determinant $(6')$ of order 3, the minors of b_1, b_2, b_3 are respectively

$$B_1 = \begin{vmatrix} a_2 & c_2 \\ a_3 & c_3 \end{vmatrix}, \qquad B_2 = \begin{vmatrix} a_1 & c_1 \\ a_3 & c_3 \end{vmatrix}, \qquad B_3 = \begin{vmatrix} a_1 & c_1 \\ a_2 & c_2 \end{vmatrix}.$$

Again, $(6')$ is the minor of d_4 in the determinant of order 4 given by (7).

90. Expansion According to the Elements of a Row or Column. In

$$(6') \qquad\qquad D = \begin{vmatrix} a_1 & b_1 & c_1 \\ a_2 & b_2 & c_2 \\ a_3 & b_3 & c_3 \end{vmatrix},$$

denote the minor of any element by the corresponding capital letter, so that b_1 has the minor B_1, b_3 has the minor B_3, etc., as in §89. We shall prove that

$$\begin{aligned} D &= a_1 A_1 - b_1 B_1 + c_1 C_1, & D &= a_1 A_1 - a_2 A_2 + a_3 A_3, \\ D &= -a_2 A_2 + b_2 B_2 - c_2 C_2, & D &= -b_1 B_1 + b_2 B_2 - b_3 B_3, \\ D &= a_3 A_3 - b_3 B_3 + c_3 C_3, & D &= c_1 C_1 - c_2 C_2 + c_3 C_3. \end{aligned}$$

The three relations at the left (or right) are expressed in words by saying that a *determinant D of the third order may be expanded according to the elements of the first, second or third row (or column)*. To obtain the expansion, we multiply each element of the row (or column) by the minor of the element, prefix the proper sign to the product, and add the signed products. The signs are alternately $+$ and $-$, as in the diagram

$$\begin{matrix} + & - & + \\ - & + & - \\ + & - & + \end{matrix}$$

For example, by expansion according to the second column,

$$\begin{vmatrix} 1 & 4 & 5 \\ 2 & 0 & 3 \\ 3 & 0 & 9 \end{vmatrix} = -4 \begin{vmatrix} 2 & 3 \\ 3 & 9 \end{vmatrix} = -4 \times 9 = -36.$$

Similarly the value of the determinant (7) of order 4 may be found by expansion according to the elements of the fourth column:

$$-d_1 \begin{vmatrix} a_2 & b_2 & c_2 \\ a_3 & b_3 & c_3 \\ a_4 & b_4 & c_4 \end{vmatrix} + d_2 \begin{vmatrix} a_1 & b_1 & c_1 \\ a_3 & b_3 & c_3 \\ a_4 & b_4 & c_4 \end{vmatrix} - d_3 \begin{vmatrix} a_1 & b_1 & c_1 \\ a_2 & b_2 & c_2 \\ a_4 & b_4 & c_4 \end{vmatrix} + d_4 \begin{vmatrix} a_1 & b_1 & c_1 \\ a_2 & b_2 & c_2 \\ a_3 & b_3 & c_3 \end{vmatrix}.$$

We shall now prove that *any determinant D of order n may be expanded according to the elements of any row or any column.*

Let E_{ij} denote the minor of e_{ij} in D, given by (8), so that E_{ij} is obtained by erasing the ith row and jth column of D.

(i) We first prove that

(10) $$D = e_{11}E_{11} - e_{21}E_{21} + e_{31}E_{31} - \cdots + (-1)^{n-1}e_{n1}E_{n1},$$

so that D may be expanded according to the elements of its first column. By (9) the terms of D having the factor e_{11} are of the form

$$(-1)^i e_{11}e_{i_2 2}\cdots e_{i_n n},$$

where $1, i_2, \ldots, i_n$ is an arrangement of $1, 2, \ldots, n$, obtained from the latter by i interchanges, so that $i_2, \ldots, i_n$ is an arrangement of $2, \ldots, n$, derived from the latter by i interchanges. After removing from each term the common factor e_{11} and adding the quotients, we obtain a sum which, by definition, is the value of the determinant E_{11} of order $n-1$. Hence the terms of D having the factor e_{11} may all be combined into $e_{11}E_{11}$, which is the first part of (10).

We next prove that the terms of D having the factor e_{21} may be combined into $-e_{21}E_{21}$, which is the second part of (10). For, if Δ be the determinant obtained from D by interchanging its first and second rows, the result just proved shows that the terms of Δ having the factor e_{21} may be combined into the product of e_{21} by the minor

$$\begin{vmatrix} e_{12} & e_{13} & \cdots & e_{1n} \\ e_{32} & e_{33} & \cdots & e_{3n} \\ \cdots\cdots\cdots\cdots\cdots \\ e_{n2} & e_{n3} & \cdots & e_{nn} \end{vmatrix}$$

of e_{21} in Δ. Now this minor is identical with the minor E_{21} of e_{21} in D. But $\Delta = -D$ (§87). Hence the terms of D having the factor e_{21} may be combined into $-e_{21}E_{21}$. Similarly, the terms of D having the factor e_{31} may be combined into $e_{31}E_{31}$, etc., as in (10).

(ii) We next prove that D may be expanded according to the elements of its kth column ($k > 1$):

$$(11) \qquad\qquad D = \sum_{j=1}^{n} (-1)^{j+k} e_{jk} E_{jk}.$$

Consider the determinant δ derived from D by moving the kth column over the earlier columns until it becomes the new first column. Since this may be done by $k - 1$ interchanges of adjacent columns, $\delta = (-1)^{k-1}D$. The minors of the elements $e_{1k}, \ldots, e_{nk}$ in the first column of δ are evidently the minors $E_{1k}, \ldots, E_{nk}$ of $e_{1k}, \ldots, e_{nk}$ in D. Hence, by (10),

$$\delta = e_{1k}E_{1k} - e_{2k}E_{2k} + \cdots + (-1)^{n-1} e_{nk}E_{nk} = \sum_{j=1}^{n} (-1)^{j+1} e_{jk} E_{jk}.$$

Thus $D = (-1)^{k-1}\delta$ has the desired value (11).

(iii) Finally, D may be expanded according to the elements of its kth row:

$$D = \sum_{j=1}^{n} (-1)^{j+k} e_{kj} E_{kj}.$$

In fact, by Case (ii), the latter is the expansion of the equal determinant D' in §85 according to the elements of its kth column.

91. Removal of Factors. *A common factor of all of the elements of the same row or same column of a determinant may be divided out of the elements and placed as a factor before the new determinant.*

In other words, if all of the elements of a row or column are divided by n, the value of the determinant is divided by n. For example,

$$\begin{vmatrix} na_1 & nb_1 \\ a_2 & b_2 \end{vmatrix} = n \begin{vmatrix} a_1 & b_1 \\ a_2 & b_2 \end{vmatrix}, \qquad \begin{vmatrix} a_1 & nb_1 & c_1 \\ a_2 & nb_2 & c_2 \\ a_3 & nb_3 & c_3 \end{vmatrix} = n \begin{vmatrix} a_1 & b_1 & c_1 \\ a_2 & b_2 & c_2 \\ a_3 & b_3 & c_3 \end{vmatrix}.$$

Proof is made by expanding the determinants according to the elements of the row or column in question and noting that the minors are the same for the two determinants. Thus the second equation is equivalent to

$$-(nb_1)B_1 + (nb_2)B_2 - (nb_3)B_3 = n(-b_1B_1 + b_2B_2 - b_3B_3),$$

where B_i denotes the minor of b_i in the final determinant.

EXERCISES

1. $\begin{vmatrix} 3a & 3b & 3c \\ 5a & 5b & 5c \\ d & e & f \end{vmatrix} = 0.$

2. $\begin{vmatrix} 2r & l & 3r \\ 2s & m & 3s \\ 2t & n & 3t \end{vmatrix} = 0.$

Expand by the shortest method and evaluate

3. $\begin{vmatrix} 2 & 7 & 3 \\ 5 & 9 & 8 \\ 0 & 3 & 0 \end{vmatrix}.$

4. $\begin{vmatrix} 5 & 7 & 0 \\ 6 & 8 & 0 \\ 3 & 9 & 4 \end{vmatrix}.$

5. $\begin{vmatrix} a & b & c & d \\ a^2 & b^2 & c^2 & d^2 \\ a^3 & b^3 & c^3 & d^3 \\ a^4 & b^4 & c^4 & d^4 \end{vmatrix} = abcd(a-b)(a-c)(a-d)(b-c)(b-d)(c-d).$

92. Sum of Determinants. *A determinant having $a_1 + q_1$, $a_2 + q_2, \ldots$ as the elements of a column is equal to the sum of the determinant having a_1, $a_2, \ldots$ as the elements of the corresponding column and the determinant having $q_1, q_2, \ldots$ as the elements of that column, while the elements of the remaining columns of each determinant are the same as in the given determinant.*

For example,

$$\begin{vmatrix} a_1 + q_1 & b_1 & c_1 \\ a_2 + q_2 & b_2 & c_2 \\ a_3 + q_3 & b_3 & c_3 \end{vmatrix} = \begin{vmatrix} a_1 & b_1 & c_1 \\ a_2 & b_2 & c_2 \\ a_3 & b_3 & c_3 \end{vmatrix} + \begin{vmatrix} q_1 & b_1 & c_1 \\ q_2 & b_2 & c_2 \\ q_3 & b_3 & c_3 \end{vmatrix}.$$

To prove the theorem we have only to expand the three determinants according to the elements of the column in question (the first column in the example) and note that the minors are the same for all three determinants. Hence $a_1 + q_1$ is multiplied by the same minor that a_1 and q_1 are multiplied by separately, and similarly for $a_2 + q_2$, etc.

The similar theorem concerning the splitting of the elements of any row into two parts is proved by expanding the three determinants according to the elements of the row in question. For example,

$$\begin{vmatrix} a+r & b+s \\ c & d \end{vmatrix} = \begin{vmatrix} a & b \\ c & d \end{vmatrix} + \begin{vmatrix} r & s \\ c & d \end{vmatrix}.$$

93. Addition of Columns or Rows.   *A determinant is not changed in value if we add to the elements of any column the products of the corresponding elements of another column by the same arbitrary number.*

Let $a_1, a_2, \ldots$ be the elements to which we add the products of the elements $b_1, b_2, \ldots$ by n. We apply §92 with $q_1 = nb_1$, $q_2 = nb_2, \ldots$. Thus the modified determinant is equal to the sum of the initial determinant and a determinant having $b_1, b_2, \ldots$ in one column and $nb_1, nb_2, \ldots$ in another column. But (§91) the latter determinant is equal to the product of n by a determinant with two columns alike and hence is zero (§88). For example,

$$\begin{vmatrix} a_1 + nb_1 & b_1 & c_1 \\ a_2 + nb_2 & b_2 & c_2 \\ a_3 + nb_3 & b_3 & c_3 \end{vmatrix} = \begin{vmatrix} a_1 & b_1 & c_1 \\ a_2 & b_2 & c_2 \\ a_3 & b_3 & c_3 \end{vmatrix} + n \begin{vmatrix} b_1 & b_1 & c_1 \\ b_2 & b_2 & c_2 \\ b_3 & b_3 & c_3 \end{vmatrix},$$

and the last determinant is zero.

Similarly, *a determinant is not changed in value if we add to the elements of any row the products of the corresponding elements of another row by the same arbitrary number.*

For example,

$$\begin{vmatrix} a + nc & b + nd \\ c & d \end{vmatrix} = \begin{vmatrix} a & b \\ c & d \end{vmatrix} + n \begin{vmatrix} c & d \\ c & d \end{vmatrix} = \begin{vmatrix} a & b \\ c & d \end{vmatrix}.$$

EXAMPLE.   Evaluate the first determinant below.

$$\begin{vmatrix} 1 & -2 & 1 \\ 1 & 2 & 3 \\ 6 & 4 & 3 \end{vmatrix} = \begin{vmatrix} 1 & 0 & 1 \\ 1 & 8 & 3 \\ 6 & 10 & 3 \end{vmatrix} = \begin{vmatrix} 0 & 0 & 1 \\ -2 & 8 & 3 \\ 3 & 10 & 3 \end{vmatrix} = \begin{vmatrix} -2 & 8 \\ 3 & 10 \end{vmatrix} = -44.$$

Solution.   First we add to the elements of the second column the products of the elements of the last column by 2. In the resulting second determinant, we add to the elements of the first column the products of the elements of the third column by -1. Finally, we expand the resulting third determinant according to the elements of its first row.

EXERCISES

1. Prove that

$$\begin{vmatrix} b + c & c + a & a + b \\ b_1 + c_1 & c_1 + a_1 & a_1 + b_1 \\ b_2 + c_2 & c_2 + a_2 & a_2 + b_2 \end{vmatrix} = 2 \begin{vmatrix} a & b & c \\ a_1 & b_1 & c_1 \\ a_2 & b_2 & c_2 \end{vmatrix}$$

By reducing to a determinant of order 3, etc., prove that

2.
$$\begin{vmatrix} 1 & 1 & 1 & 1 \\ a & b & c & d \\ a^2 & b^2 & c^2 & d^2 \\ a^3 & b^3 & c^3 & d^3 \end{vmatrix} = (a-b)(a-c)(a-d)(b-c)(b-d)(c-d).$$

3.
$$\begin{vmatrix} 2 & -1 & 3 & -2 \\ 1 & 7 & 1 & -1 \\ 3 & 5 & -5 & 3 \\ 4 & -3 & 2 & -1 \end{vmatrix} = -42.$$

4.
$$\begin{vmatrix} 1 & 1 & 1 & 1 \\ 1 & 2 & 3 & 4 \\ 1 & 3 & 6 & 10 \\ 1 & 4 & 10 & 20 \end{vmatrix} = 1.$$

94. System of n Linear Equations in n Unknowns with $D \neq 0$. In

$$(12) \qquad \begin{aligned} a_{11}x_1 + a_{12}x_2 + \cdots + a_{1n}x_n &= k_1, \\ &\cdots\cdots\cdots\cdots\cdots\cdots\cdots \\ a_{n1}x_1 + a_{n2}x_2 + \cdots + a_{nn}x_n &= k_n, \end{aligned}$$

let D denote the determinant of the coefficients of the n unknowns:

$$D = \begin{vmatrix} a_{11} & a_{12} & \cdots & a_{1n} \\ \cdots & \cdots & \cdots & \cdots \\ a_{n1} & a_{n2} & \cdots & a_{nn} \end{vmatrix}.$$

Then

$$Dx_1 = \begin{vmatrix} a_{11}x_1 & a_{12} & \cdots & a_{1n} \\ \cdots & \cdots & \cdots & \cdots \\ a_{n1}x_1 & a_{n2} & \cdots & a_{nn} \end{vmatrix} = \begin{vmatrix} a_{11}x_1 + a_{12}x_2 + \cdots + a_{1n}x_n & a_{12} & \cdots & a_{1n} \\ \cdots & \cdots & \cdots & \cdots \\ a_{n1}x_1 + a_{n2}x_2 + \cdots + a_{nn}x_n & a_{n2} & \cdots & a_{nn} \end{vmatrix},$$

where the second determinant was derived from the first by adding to the elements of the first column the products of the corresponding elements of the second column by x_2, etc., and finally the products of the elements of the last column by x_n. The elements of the new first column are equal to $k_1, \ldots, k_n$ by (12). In this manner, we find that

$$(13) \qquad Dx_1 = K_1, \qquad Dx_2 = K_2, \qquad \ldots, \qquad Dx_n = K_n,$$

in which K_i is derived from D by substituting $k_1, \ldots, k_n$ for the elements $a_{1i}, \ldots, a_{ni}$ of the ith column of D, whence

$$K_1 = \begin{vmatrix} k_1 & a_{12} & \cdots & a_{1n} \\ \cdots & \cdots & \cdots & \cdots \\ k_n & a_{n2} & \cdots & a_{nn} \end{vmatrix}, \ldots \qquad K_n = \begin{vmatrix} a_{11} & \cdots & a_{1n-1} & k_1 \\ \cdots & \cdots & \cdots & \cdots \\ a_{n1} & \cdots & a_{nn-1} & k_n \end{vmatrix}.$$

If $D \neq 0$, the unique values of $x_1, \ldots, x_n$ determined by division from (13) actually satisfy equations (12). For instance, the first equation is satisfied since

$$k_1 D - a_{11} K_1 - a_{12} K_2 - \cdots - a_{1n} K_n = \begin{vmatrix} k_1 & a_{11} & a_{12} & \cdots & a_{1n} \\ k_1 & a_{11} & a_{12} & \cdots & a_{1n} \\ k_2 & a_{21} & a_{22} & \cdots & a_{2n} \\ \cdots & \cdots & \cdots & \cdots & \cdots \\ k_n & a_{n1} & a_{n2} & \cdots & a_{nn} \end{vmatrix},$$

as shown by expansion according to the elements of the first row; and the determinant is zero, having two rows alike.

THEOREM. *If D denotes the determinant of the coefficients of the n unknowns in a system of n linear equations, the product of D by any one of the unknowns is equal to the determinant obtained from D by substituting the known terms in place of the coefficients of that unknown. If $D \neq 0$, we obtain the unique values of the unknowns by division by D.*

We have therefore given a complete proof of the results stated and illustrated in §80, §81. Another proof is suggested in Ex. 7 below. The theorem was discovered by induction in 1750 by G. Cramer.

EXERCISES

Solve by determinants the following systems of equations (reducing each determinant to one having zero as the value of every element but one in a row or column, as in the example in §93).

1. $x + y + z = 11,$
    $2x - 6y - z = 0,$
    $3x + 4y + 2z = 0.$

2. $x + y + z = 0,$
    $x + 2y + 3z = -1,$
    $x + 3y + 6z = 0.$

3. $x - 2y + z = 12,$
    $x + 2y + 3z = 48,$
    $6x + 4y + 3z = 84.$

4. $3x - 2y = 7,$
    $3y - 2z = 6,$
    $3z - 2x = -1.$

5. $x + y + z + w = 1,$
    $x + 2y + 3z + 4w = 11,$
    $x + 3y + 6z + 10w = 26,$
    $x + 4y + 10z + 20w = 47.$

6. $2x - y + 3z - 2w = 4,$
    $x + 7y + z - w = 2,$
    $3x + 5y - 5z + 3w = 0,$
    $4x - 3y + 2z - w = 5.$

7. Prove the first relation (13) by multiplying the members of the first equation (12) by A_{11}, those of the second equation by $-A_{21}, \ldots$, those of the nth equation by $(-1)^{n-1}A_{n1}$, and adding, where A_{ij} by denotes the minor of a_{ij} in D. Hint: The resulting coefficient of x_2 is the expansion, according to the elements of its first column, of a determinant derived from D by replacing a_{11} by $a_{12}, \ldots, a_{n1}$ by a_{n2}.

95. Rank of a Determinant.

If we erase from a determinant D of order n all but r rows and all but r columns, we obtain a determinant of order r called an *r-rowed minor of D*. In particular, any element is regarded as a one-rowed minor, and D itself is regarded as an n-rowed minor.

If a determinant D of order n is not zero, it is said to be of *rank n*. If, for $0 < r < n$, some r-rowed minor of D is not zero, while every $(r + 1)$-rowed minor is zero, D is said to be of *rank r*. It is said to be of rank zero if every element is zero.

For example, a determinant D of order 3 is of rank 3 if $D \neq 0$; of rank 2 if $D = 0$, but some two-rowed minor is not zero; of rank 1 if every two-rowed minor is zero, but some element is not zero. Again, every three-rowed minor of

$$\begin{vmatrix} a & b & c & d \\ e & f & g & h \\ a & b & c & d \\ e & f & g & h \end{vmatrix}$$

is zero since two pairs of its rows are alike. Hence it is of rank 2 if some two-rowed minor is not zero. But it is of rank 1 if a, b, c, d are not all zero and are proportional to e, f, g, h, since all two-rowed minors are then zero.

96. System of n Linear Equations in n Unknowns with $D = 0$.

We shall now discuss the equations (12) for the troublesome case (previously ignored) in which the determinant D of the coefficients of the unknowns is zero. In view of (13), the given equations are evidently inconsistent if any one of the determinants $K_1, \ldots, K_n$ is not zero. But if D and these K's are all zero, our former results (13) give us no information concerning the unknowns x_i, and we resort to the following

THEOREM.   *Let the determinant D of the coefficients of the unknowns in equations (12) be of rank r, $r < n$. If the determinants K obtained from the $(r + 1)$-rowed minors of D by replacing the elements of any column by the corresponding known terms k_i are not all zero, the equations are inconsistent. But if these determinants K are all zero, the r equations involving the elements of a non-vanishing r-rowed minor of D determine uniquely r of the unknowns as linear functions of the remaining $n - r$ unknowns, which are independent* *variables, and the expressions for these r unknowns satisfy also the remaining $n - r$ equations.*

Consider for example the three equations (4) in the unknowns x, y, z. Five cases arise:

(α) D of rank 3, i.e., $D \neq 0$.

(β) D of rank 2 (i.e., $D = 0$, but some two-rowed minor $\neq 0$), and

$$K_1 = \begin{vmatrix} k_1 & b_1 & c_1 \\ k_2 & b_2 & c_2 \\ k_3 & b_3 & c_3 \end{vmatrix}, \quad K_2 = \begin{vmatrix} a_1 & k_1 & c_1 \\ a_2 & k_2 & c_2 \\ a_3 & k_3 & c_3 \end{vmatrix}, \quad K_3 = \begin{vmatrix} a_1 & b_1 & k_1 \\ a_2 & b_2 & k_2 \\ a_3 & b_3 & k_3 \end{vmatrix}$$

not all zero.

(γ) D of rank 2 and K_1, K_2, K_3 all zero.

(δ) D of rank 1 (i.e., every two-rowed minor $= 0$, but some element $\neq 0$), and

$$\begin{vmatrix} a_i & k_i \\ a_j & k_j \end{vmatrix}, \quad \begin{vmatrix} b_i & k_i \\ b_j & k_j \end{vmatrix}, \quad \begin{vmatrix} c_i & k_i \\ c_j & k_j \end{vmatrix} \quad (i, j \text{ chosen from 1, 2, 3})$$

not all zero; there are nine such determinants K.

(ϵ) D of rank 1, and all nine of the two-rowed determinants K zero.

In case (α) the equations have a single set of solutions (§94). In cases (β) and (δ) there is no set of solutions. For (β) the proof follows from (13). In case (γ) one of the equations is a linear combination of the other two; for example, if $a_1 b_2 - a_2 b_1 \neq 0$, the first two equations determine x and y as linear functions of z (as shown by transposing the terms in z and solving the resulting equations for x and y), and the resulting values of x and y satisfy the third equation identically as to z. Finally, in case (ϵ), two of the equations are obtained by multiplying the remaining one by constants.

The reader acquainted with the elements of solid analytic geometry will see that the planes represented by the three equations have the following relations:

(α) The three planes intersect in a single point.

(β) Two of the planes intersect in a line parallel to the third plane.

(γ) The three planes intersect in a common line.

(δ) The three planes are parallel and not all coincident.

(ϵ) The three planes coincide.

The remarks preceding our theorem furnish an illustration (the case $r = n - 1$) of the following

LEMMA 1.   *If every $(r+1)$-rowed minor M formed from certain $r+1$ rows of D is zero, the corresponding $r + 1$ equations (12) are inconsistent provided there is a non-vanishing determinant K formed from any M by replacing the elements of any column by the corresponding known terms k_i.*

For concreteness,[1] let the rows in question be the first $r + 1$ and let

$$K = \begin{vmatrix} a_{11} & \cdots & a_{1r} & k_1 \\ \hdotsfor{4} \\ a_{r+11} & \cdots & a_{r+1r} & k_{r+1} \end{vmatrix} \neq 0.$$

Let $d_1, \ldots, d_{r+1}$ be the minors of $k_1, \ldots, k_{r+1}$ in K. Multiply the first $r + 1$ equations (12) by $d_1, -d_2, \ldots, (-1)^r d_{r+1}$, respectively, and add. The right member of the resulting equation is the expansion of $\pm K$. The coefficient of x_s is the expansion of

$$\pm \begin{vmatrix} a_{11} & \cdots & a_{1r} & a_{1s} \\ \hdotsfor{4} \\ a_{r+11} & \cdots & a_{r+1r} & a_{r+1s} \end{vmatrix}$$

and is zero, being an M if $s > r$, and having two columns identical if $s \leq r$. Hence $0 = \pm K$. Thus if $K \neq 0$, the equations are inconsistent.

LEMMA 2.   *If all of the determinants M and K in Lemma 1 are zero, but an r-rowed minor of an M is not zero, one of the corresponding $r+1$ equations is a linear combination of the remaining r equations.*

As before let the $r + 1$ rows in question be the first $r + 1$. Let the non-vanishing r-rowed minor be

$$(14) \qquad\qquad d_{r+1} = \begin{vmatrix} a_{11} & \cdots & a_{1r} \\ \hdotsfor{3} \\ a_{r1} & \cdots & a_{rr} \end{vmatrix} \neq 0.$$

Let the functions obtained by transposing the terms k_i in (12) be

$$L_i \equiv a_{i1}x_1 + a_{i2}x_2 + \cdots + a_{in}x_n - k_i.$$

By the multiplication made in the proof of Lemma 1,

$$d_1 L_1 - d_2 L_2 + \cdots + (-1)^r d_{r+1} L_{r+1} = \mp K = 0.$$

Hence L_{r+1} is a linear combination of $L_1, \ldots, L_r$.

---

[1] All other cases may be reduced to this one by rearranging the n equations and relabelling the unknowns (replacing x_3 by the new x_1, for example).

The first part of the theorem is true by Lemma 1. The second part is readily proved by means of Lemma 2. Let (14) be the non-vanishing r-rowed minor of D. For $s > r$, the sth equation is a linear combination of the first r equations, and hence is satisfied by any set of solutions of the latter. In the latter transpose the terms involving $x_{r+1}, \ldots, x_n$. Since the determinant of the coefficients of $x_1, \ldots, x_r$ is not zero, §94 shows that $x_1, \ldots, x_r$ are uniquely determined linear functions of $x_{r+1}, \ldots, x_n$ (which enter from the new right members).

EXERCISES

Apply the theorem to the following four systems of equations and check the conclusions:

1.
$$2x + y + 3z = 1,$$
$$4x + 2y - z = -3,$$
$$2x + y - 4z = -4.$$

2.
$$2x + y + 3z = 1,$$
$$4x + 2y - z = 3,$$
$$2x + y - 4z = 4.$$

3.
$$x - 3y + 4z = 1,$$
$$4x - 12y + 16z = 3,$$
$$3x - 9y + 12z = 3.$$

4.
$$x - 3y + 4z = 1,$$
$$4x - 12y + 16z = 4,$$
$$3x - 9y + 12z = 3.$$

5. Discuss the system

$$ax + y + z = a - 3,$$
$$x + ay + z = -2,$$
$$x + y + az = -2,$$

when (*i*) $a = 1$; (*ii*) $a = -2$; (*iii*) $a \neq 1, -2$, obtaining the simplest forms of the unknowns.

6. Discuss the system

$$x + y + z = 1,$$
$$ax + by + cz = k,$$
$$a^2x + b^2y + c^2z = k^2,$$

when (*i*) a, b, c are distinct; (*ii*) $a = b \neq c$; (*iii*) $a = b = c$.

97. Homogeneous Linear Equations. When the known terms k_1, $\ldots, k_n$ in (12) are all zero, the equations are called *homogeneous*. The determinants K are now all zero, so that the n homogeneous equations are never inconsistent. This is also evident from the fact that they have the set of solutions $x_1 = 0, \ldots, x_n = 0$. By (13), there is no further set of solutions if $D \neq 0$. If $D = 0$, there are further sets of solutions. This is shown by the theorem of §96 which now takes the following simpler form.

If the determinant D of the coefficients of n linear homogeneous equations in n unknowns is of rank r, r < n, the r equations involving the elements of a non-vanishing r-rowed minor of D determine uniquely r of the unknowns as linear functions of the remaining n − r unknowns, which are independent variables, and the expressions for these r unknowns satisfy also the remaining n − r equations.

The particular case mentioned is the much used theorem:

A necessary and sufficient condition that n linear homogeneous equations in n unknowns shall have a set of solutions, other than the trivial one in which each unknown is zero, is that the determinant of the coefficients be zero.

EXERCISES

Discuss the following systems of equations:

1. $x + y + 3z = 0,$
    $x + 2y + 2z = 0,$
    $x + 5y - z = 0.$

2. $2x - y + 4z = 0,$
    $x + 3y - 2z = 0,$
    $x - 11y + 14z = 0.$

3. $x - 3y + 4z = 0,$
    $4x - 12y + 16z = 0,$
    $3x - 9y + 12z = 0.$

4. $6x + 4y + 3z - 84w = 0,$
    $x + 2y + 3z - 48w = 0,$
    $x - 2y + z - 12w = 0,$
    $4x + 4y - z - 24w = 0.$

5. $2x + 3y - 4z + 5w = 0,$
    $3x + 5y - z + 2w = 0,$
    $7x + 11y - 9z + 12w = 0,$
    $3x + 4y - 11z + 13w = 0.$

98. System of m Linear Equations in n Unknowns. The case $m < n$ may be treated by means of the lemmas in §96. If $m > n$, we select any n of the equations and apply to them the theorems of §§94, 96. If they are found to be inconsistent, the entire system is evidently inconsistent. But if the n equations are consistent, and if r is the rank of the determinant of their coefficients, we obtain r of the unknowns expressed as linear functions of the remaining $n - r$ unknowns. Substituting these values of these r unknowns in the remaining equations, we obtain a system of $m - n$ linear equations in $n - r$ unknowns. Treating this system in the same manner, we ultimately either find that the proposed m equations are consistent and obtain the general set of solutions, or find that they are inconsistent. To decide in advance whether the former or latter of these cases will arise, we have only to find the maximum order r of a non-vanishing r-rowed determinant formed from the coefficients of the unknowns, taken in the regular order in which they occur in the equations, and ascertain whether or not the corresponding $(r + 1)$-rowed determinants K, formed as in §96, are all zero.

The last result may be expressed simply by employing the terminology of matrices. The system of coefficients of the unknowns in any set of linear equations

$$(15) \qquad \begin{array}{c} a_{11}x_1 + \cdots + a_{1n}x_n = k_1, \\ \cdots\cdots\cdots\cdots\cdots\cdots\cdots\cdots \\ a_{m1}x_1 + \cdots + a_{mn}x_n = k_m, \end{array}$$

arranged as they occur in the equations, is called the *matrix* of the coefficients, and is denoted by

$$A = \begin{pmatrix} a_{11} & a_{12} & \cdots & a_{1n} \\ \cdots\cdots\cdots\cdots\cdots\cdots \\ a_{m1} & a_{m2} & \cdots & a_{mn} \end{pmatrix}.$$

By annexing the column composed of the known terms k_i we obtain the so-called *augmented matrix*

$$B = \begin{pmatrix} a_{11} & a_{12} & \cdots & a_{1n} & k_1 \\ \cdots\cdots\cdots\cdots\cdots\cdots\cdots \\ a_{m1} & a_{m2} & \cdots & a_{mn} & k_m \end{pmatrix}.$$

The definitions of an r-rowed minor (determinant) of a matrix and of the rank of a matrix are entirely analogous to the definitions in §95.

In view of Lemma 1 in §96, our equations (15) are inconsistent if B is of rank $r + 1$ and A is of rank $\leqq r$. By Lemma 2, if A and B are both of rank r, all of our equations are linear combinations of r of them. Noting also that the rank r of A cannot exceed the rank of B, since every minor of A is a minor of B, and hence a non-vanishing r-rowed minor of A is a minor of B, so that the rank of B is not less than r, we have the following

THEOREM.   *A system of m linear equations in n unknowns is consistent if and only if the rank of the matrix of the coefficients of the unknowns is equal to the rank of the augmented matrix. If the rank of both matrices is r, certain r of the equations determine uniquely r of the unknowns as linear functions of the remaining $n - r$ unknowns, which are independent variables, and the expressions for these r unknowns satisfy also the remaining $m - r$ equations.*

When $m = n + 1$, B has an m-rowed minor called the *determinant of the square matrix B*. If this determinant is not zero, B is of rank m. Since A has no m-rowed minor, its rank is less than m. Hence we obtain the

COROLLARY.   *Any system of $n + 1$ linear equations in n unknowns is inconsistent if the determinant of the augmented matrix is not zero.*

EXERCISES

Discuss the following systems of equations:

1.
$$2x + y + 3z = 1,$$
$$4x + 2y - z = -3,$$
$$2x + y - 4z = -4,$$
$$10x + 5y - 6z = -10.$$

2.
$$2x - y + 3z = 2,$$
$$x + 7y + z = 1,$$
$$3x + 5y - 5z = a,$$
$$4x - 3y + 2z = 1.$$

3.
$$4x - y + z = 5,$$
$$2x - 3y + 5z = 1,$$
$$x + y - 2z = 2,$$
$$5x - z = 2.$$

4.
$$4x - 5y = 2,$$
$$2x + 3y = 12,$$
$$10x - 7y = 16.$$

5. Prove the Corollary by multiplying the known terms by $x_{n+1} = 1$ and applying §97 with n replaced by $n + 1$.

6. Prove that if the matrix of the coefficients of any system of linear homogeneous equations in n unknowns is of rank r, the values of certain $n-r$ of the unknowns may be assigned at pleasure and the others will then be uniquely determined and satisfy all of the equations.

99. Complementary Minors.   The determinant

(16)
$$D = \begin{vmatrix} a_1 & b_1 & c_1 & d_1 \\ a_2 & b_2 & c_2 & d_2 \\ a_3 & b_3 & c_3 & d_3 \\ a_4 & b_4 & c_4 & d_4 \end{vmatrix}$$

is said to have the *two-rowed complementary minors*

$$M = \begin{vmatrix} a_1 & b_1 \\ a_3 & b_3 \end{vmatrix}, \qquad M' = \begin{vmatrix} c_2 & d_2 \\ c_4 & d_4 \end{vmatrix},$$

since either is obtained by erasing from D all the rows and columns having an element which occurs in the other.

In general, if we erase from a determinant D of order n all but r rows and all but r columns, we obtain a determinant M of order r called an r-rowed minor of D. But if we had erased from D the r rows and r columns previously kept, we would have obtained an $(n - r)$-rowed minor of D called the *minor complementary to M*. In particular, any element is regarded as a one-rowed minor and is complementary to its minor (of order $n - 1$).

100. Laplace's Development by Columns. *Any determinant D is equal to the sum of all the signed products $\pm M M'$, where M is an r-rowed minor having its elements in the first r columns of D, and M' is the minor complementary to M, while the sign is $+$ or $-$ according as an even or odd number of interchanges of rows of D will bring M into the position occupied by the minor M_1 whose elements lie in the first r rows and first r columns of D.*

For $r = 1$, this development becomes the known expansion of D according to the elements of the first column (§90); here $M_1 = e_{11}$.

If $r = 2$ and D is the determinant (16),

$$D = \begin{vmatrix} a_1 & b_1 \\ a_2 & b_2 \end{vmatrix} \cdot \begin{vmatrix} c_3 & d_3 \\ c_4 & d_4 \end{vmatrix} - \begin{vmatrix} a_1 & b_1 \\ a_3 & b_3 \end{vmatrix} \cdot \begin{vmatrix} c_2 & d_2 \\ c_4 & d_4 \end{vmatrix} + \begin{vmatrix} a_1 & b_1 \\ a_4 & b_4 \end{vmatrix} \cdot \begin{vmatrix} c_2 & d_2 \\ c_3 & d_3 \end{vmatrix}$$

$$+ \begin{vmatrix} a_2 & b_2 \\ a_3 & b_3 \end{vmatrix} \cdot \begin{vmatrix} c_1 & d_1 \\ c_4 & d_4 \end{vmatrix} - \begin{vmatrix} a_2 & b_2 \\ a_4 & b_4 \end{vmatrix} \cdot \begin{vmatrix} c_1 & d_1 \\ c_3 & d_3 \end{vmatrix} + \begin{vmatrix} a_3 & b_3 \\ a_4 & b_4 \end{vmatrix} \cdot \begin{vmatrix} c_1 & d_1 \\ c_2 & d_2 \end{vmatrix}.$$

The first product in the development is $M_1 M_1'$; the second product is $-M M'$ (in the notations of §99), and the sign is minus since the interchange of the second and third rows of D brings this M into the position of M_1. The sign of the third product in the development is plus since two interchanges of rows of D bring the first factor into the position of M_1.

If D is the determinant (8), then

$$M_1 = \begin{vmatrix} e_{11} & \cdots & e_{1r} \\ \cdots\cdots\cdots \\ e_{r1} & \cdots & e_{rr} \end{vmatrix}, \qquad M_1' = \begin{vmatrix} e_{r+1\,r+1} & \cdots & e_{r+1\,n} \\ \cdots\cdots\cdots\cdots\cdots \\ e_{n\,r+1} & \cdots & e_{nn} \end{vmatrix}.$$

Any term of the product $M_1 M_1'$ is of the type

$$(17) \qquad (-1)^i e_{i_1 1} e_{i_2 2} \cdots e_{i_r r} \cdot (-1)^j e_{i_{r+1} r+1} \cdots e_{i_n n},$$

where $i_1, \ldots, i_r$ is an arrangement of $1, \ldots, r$ derived from $1, \ldots, r$ by i interchanges, while $i_{r+1}, \ldots, i_n$ is an arrangement of $r+1, \ldots, n$ derived by j interchanges. Hence $i_1, \ldots, i_n$ is an arrangement of $1, \ldots, n$ derived by $i+j$ interchanges, so that the product (17) is a term of D with the proper sign.

It now follows from §87 that any term of any of the products $\pm MM'$ mentioned in the theorem is a term of D. Clearly we do not obtain twice in this manner the same term of D.

Conversely, any term t of D occurs in one of the products $\pm MM'$. Indeed, t contains as factors r elements from the first r columns of D, no two being in the same row, and the product of these is, except perhaps as to sign, a term of some minor M. Thus t is a term of MM' or of $-MM'$. In view of the earlier discussion, the sign of t is that of the corresponding term in $\pm MM'$, where the latter sign is given by the theorem.

101. Laplace's Development by Rows. There is a Laplace development of D in which the r-rowed minors M have their elements in the first r rows of D, instead of in the first r columns as in §100. To prove this, we have only to apply §100 to the equal determinant obtained by interchanging the rows and columns of D.

There are more general (but less used) Laplace developments in which the r-rowed minors M have their elements in any chosen r columns (or rows) of D. It is simpler to apply the earlier developments to the determinant $\pm D$ having the elements of the chosen r columns (or rows) in the new first r columns (or rows).

EXERCISES

1. Prove that

$$\begin{vmatrix} a & b & c & d \\ e & f & g & h \\ 0 & 0 & j & k \\ 0 & 0 & l & m \end{vmatrix} = \begin{vmatrix} a & b \\ e & f \end{vmatrix} \cdot \begin{vmatrix} j & k \\ l & m \end{vmatrix}.$$

2. By employing 2-rowed minors from the first two rows, show that

$$\frac{1}{2}\begin{vmatrix} a & b & c & d \\ e & f & g & h \\ a & b & c & d \\ e & f & g & h \end{vmatrix} = \begin{vmatrix} a & b \\ e & f \end{vmatrix} \cdot \begin{vmatrix} c & d \\ g & h \end{vmatrix} - \begin{vmatrix} a & c \\ e & g \end{vmatrix} \cdot \begin{vmatrix} b & d \\ f & h \end{vmatrix} + \begin{vmatrix} a & d \\ e & h \end{vmatrix} \cdot \begin{vmatrix} b & c \\ f & g \end{vmatrix} = 0.$$

3. By employing 2-rowed minors from the first two columns of the 4-rowed determinant in Ex. 2, show that the products in Laplace's development cancel.

102. Product of Determinants. *The product of two determinants of the same order is equal to a determinant of like order in which the element of the rth row and cth column is the sum of the products of the elements of the rth row of the first determinant by the corresponding elements of the cth column of the second determinant.*

For example,

(18)
$$\begin{vmatrix} a & b \\ c & d \end{vmatrix} \cdot \begin{vmatrix} e & f \\ g & h \end{vmatrix} = \begin{vmatrix} ae + bg & af + bh \\ ce + dg & cf + dh \end{vmatrix}.$$

While for brevity we shall give the proof for determinants of order 3, the method is seen to apply to determinants of any order. By Laplace's development with $r = 3$ (§101), we have

(19)
$$\begin{vmatrix} a_1 & b_1 & c_1 & 0 & 0 & 0 \\ a_2 & b_2 & c_2 & 0 & 0 & 0 \\ a_3 & b_3 & c_3 & 0 & 0 & 0 \\ -1 & 0 & 0 & e_1 & f_1 & g_1 \\ 0 & -1 & 0 & e_2 & f_2 & g_2 \\ 0 & 0 & -1 & e_3 & f_3 & g_3 \end{vmatrix} = \begin{vmatrix} a_1 & b_1 & c_1 \\ a_2 & b_2 & c_2 \\ a_3 & b_3 & c_3 \end{vmatrix} \cdot \begin{vmatrix} e_1 & f_1 & g_1 \\ e_2 & f_2 & g_2 \\ e_3 & f_3 & g_3 \end{vmatrix}.$$

In the determinant of order 6, add to the elements of the fourth, fifth, and sixth columns the products of the elements of the first column by e_1, f_1, g_1, respectively (and hence introduce zeros in place of the former elements e_1, f_1, g_1). Next, add to the elements of the fourth, fifth, and sixth columns the products of the elements of the second column by e_2, f_2, g_2, respectively. Finally, add to the elements of the fourth, fifth, and sixth columns the products of the elements of the third column by e_3, f_3, g_3, respectively. The new determinant is

$$\begin{vmatrix} a_1 & b_1 & c_1 & a_1e_1 + b_1e_2 + c_1e_3 & a_1f_1 + b_1f_2 + c_1f_3 & a_1g_1 + b_1g_2 + c_1g_3 \\ a_2 & b_2 & c_2 & a_2e_1 + b_2e_2 + c_2e_3 & a_2f_1 + b_2f_2 + c_2f_3 & a_2g_1 + b_2g_2 + c_2g_3 \\ a_3 & b_3 & c_3 & a_3e_1 + b_3e_2 + c_3e_3 & a_3f_1 + b_3f_2 + c_3f_3 & a_3g_1 + b_3g_2 + c_3g_3 \\ -1 & 0 & 0 & 0 & 0 & 0 \\ 0 & -1 & 0 & 0 & 0 & 0 \\ 0 & 0 & -1 & 0 & 0 & 0 \end{vmatrix}.$$

By Laplace's development (or by expansion according to the elements of the last row, etc.), this is equal to the 3-rowed minor whose elements are the long sums. Hence this minor is equal to the product in the right member of (19).

EXERCISES

1. Prove (18) by means of §92.

2. Prove that, if $s_i = \alpha^i + \beta^i + \gamma^i$,

$$\begin{vmatrix} 1 & 1 & 1 \\ \alpha & \beta & \gamma \\ \alpha^2 & \beta^2 & \gamma^2 \end{vmatrix} \cdot \begin{vmatrix} 1 & \alpha & \alpha^2 \\ 1 & \beta & \beta^2 \\ 1 & \gamma & \gamma^2 \end{vmatrix} = \begin{vmatrix} 3 & s_1 & s_2 \\ s_1 & s_2 & s_3 \\ s_2 & s_3 & s_4 \end{vmatrix}.$$

3. If A_i, B_i, C_i are the minors of a_i, b_i, c_i in the determinant D defined by the second factor below, prove that

$$\begin{vmatrix} A_1 & -A_2 & A_3 \\ -B_1 & B_2 & -B_3 \\ C_1 & -C_2 & C_3 \end{vmatrix} \cdot \begin{vmatrix} a_1 & b_1 & c_1 \\ a_2 & b_2 & c_2 \\ a_3 & b_3 & c_3 \end{vmatrix} = \begin{vmatrix} D & 0 & 0 \\ 0 & D & 0 \\ 0 & 0 & D \end{vmatrix}.$$

Hence the first factor is equal to D^2 if $D \neq 0$.

4. Express $(a^2 + b^2 + c^2 + d^2)(e^2 + f^2 + g^2 + h^2)$ as a sum of four squares by writing

$$\begin{vmatrix} a + bi & c + di \\ -c + di & a - bi \end{vmatrix} \cdot \begin{vmatrix} e + fi & g + hi \\ -g + hi & e - fi \end{vmatrix}$$

as a determinant of order 2 similar to each factor. Hint: If k' denotes the conjugate of the complex number k, each of the three determinants is of the form

$$\begin{vmatrix} k & l \\ -l' & k' \end{vmatrix}.$$

MISCELLANEOUS EXERCISES

1. Solve

$$ax + by + cz = k,$$
$$a^2 x + b^2 y + c^2 z = k^2,$$
$$a^4 x + b^4 y + c^4 z = k^4$$

by determinants for x, treating all cases.

2. In three linear homogeneous equations in four unknowns, prove that the values of the unknowns are proportional to four determinants of order 3 formed from the coefficients.

Factor the following determinants:

3. $\begin{vmatrix} 1 & a & bc \\ 1 & b & ca \\ 1 & c & ab \end{vmatrix}$.

4. $\begin{vmatrix} x & x^2 & yz \\ y & y^2 & xz \\ z & z^2 & xy \end{vmatrix} = \begin{vmatrix} x^2 & x^3 & 1 \\ y^2 & y^3 & 1 \\ z^2 & z^3 & 1 \end{vmatrix}$.

5.

$$\begin{vmatrix} a & b & c \\ c & a & b \\ b & c & a \end{vmatrix} = (a+b+c)(a+b\omega+c\omega^2)(a+b\omega^2+c\omega),$$

where ω is an imaginary cube root of unity.

6. $\begin{vmatrix} a & b & c & d \\ b & a & d & c \\ c & d & a & b \\ d & c & b & a \end{vmatrix}$.

7. $\begin{vmatrix} a & b & c & d \\ d & a & b & c \\ c & d & a & b \\ b & c & d & a \end{vmatrix}$.

8. If the points $(x_1, y_1), \ldots, (x_4, y_4)$ lie on a circle, prove that

$$\begin{vmatrix} x_1^2 + y_1^2 & x_1 & y_1 & 1 \\ \cdots\cdots\cdots\cdots\cdots & & & \\ x_4^2 + y_4^2 & x_4 & y_4 & 1 \end{vmatrix} = 0.$$

9. Prove that

$$\begin{vmatrix} aa' + bb' + cc' & ea' + fb' + gc' \\ ae' + bf' + cg' & ee' + ff' + gg' \end{vmatrix}$$
$$= \begin{vmatrix} a & b \\ e & f \end{vmatrix} \cdot \begin{vmatrix} a' & b' \\ e' & f' \end{vmatrix} + \begin{vmatrix} a & c \\ e & g \end{vmatrix} \cdot \begin{vmatrix} a' & c' \\ e' & g' \end{vmatrix} + \begin{vmatrix} b & c \\ f & g \end{vmatrix} \cdot \begin{vmatrix} b' & c' \\ f' & g' \end{vmatrix}.$$

10. Prove that the cubic equation

$$D(x) \equiv \begin{vmatrix} a - x & b & c \\ b & f - x & g \\ c & g & h - x \end{vmatrix} = 0$$

has only real roots. Hints:

$$D(x) \cdot D(-x) = \begin{vmatrix} a^2 + b^2 + c^2 - x^2 & ab + bf + cg & ac + bg + ch \\ ab + bf + cg & b^2 + f^2 + g^2 - x^2 & bc + fg + gh \\ ac + bg + ch & bc + fg + gh & c^2 + g^2 + h^2 - x^2 \end{vmatrix}$$
$$= -x^6 + x^4(a^2 + f^2 + h^2 + 2b^2 + 2c^2 + 2g^2) - x^2(D_1 + D_2 + D_3) + D^2(0),$$

where D_3 denotes the first determinant in Ex. 9 with all accents removed and with $e = b$, while D_1 and D_2 are analogous minors of elements in the main diagonal of the present determinant of order 3 with $x = 0$. Hence the coefficient of $-x^2$ is a sum of squares. Since the function of degree 6 is not zero for a negative value of x^2,

$D(x) = 0$ has no purely imaginary root. If it had an imaginary root $r + si$, then $D(x + r) = 0$ would have a purely imaginary root si. But $D(x + r)$ is of the form $D(x)$ with a, f, h replaced by $a - r$, $f - r$, $h - r$. Hence $D(x) = 0$ has only real roots. The method is applicable to such determinants of order n.

11. If $a_1, \ldots, a_n$ are distinct, solve the system of equations

$$\frac{x_1}{k_i - a_1} + \frac{x_2}{k_i - a_2} + \cdots + \frac{x_n}{k_i - a_n} = 1 \qquad (i = 1, \ldots, n).$$

Hint: Regard $k_1, \ldots, k_n$ as the roots of an equation of degree n in k formed from the typical one above by substituting k for k_i and clearing of fractions; write $k = a_j - t$, and consider the product of the roots of $t^n + \cdots = 0$. Hence find x_j.

12. Solve the equation

$$\begin{vmatrix} a + x & x & x \\ x & b + x & x \\ x & x & c + x \end{vmatrix} = 0.$$

CHAPTER IX

SYMMETRIC FUNCTIONS

103. Sigma Functions, Elementary Symmetric Functions. A rational function of the independent variables $x_1, x_2, \ldots, x_n$ is said to be *symmetric* in them if it is unaltered by the interchange of any two of the variables. For example,

$$x_1^2 + x_2^2 + x_3^2 + 4x_1 + 4x_2 + 4x_3$$

is a symmetric polynomial in x_1, x_2, x_3; the sum of the first three terms is denoted by Σx_1^2 and the sum of the last three by $4\Sigma x_1$. In general, if t is a rational function of $x_1, \ldots, x_n$, Σt denotes the sum of t and all of the distinct functions obtained from t by permutations of the variables; such a Σ-function (read *sigma function*) is symmetric in $x_1, \ldots, x_n$.

For example, if there are three independent variables α, β, γ,

$$\Sigma\alpha\beta = \alpha\beta + \alpha\gamma + \beta\gamma, \qquad \Sigma\alpha^2\beta = \alpha^2\beta + \alpha\beta^2 + \alpha^2\gamma + \alpha\gamma^2 + \beta^2\gamma + \beta\gamma^2,$$

$$\Sigma\frac{1}{\alpha} = \frac{1}{\alpha} + \frac{1}{\beta} + \frac{1}{\gamma}, \qquad \Sigma\frac{\beta}{\alpha} = \frac{\beta}{\alpha} + \frac{\alpha}{\beta} + \frac{\beta}{\gamma} + \frac{\gamma}{\beta} + \frac{\alpha}{\gamma} + \frac{\gamma}{\alpha},$$

$$\Sigma\frac{\alpha^2 + \beta^2}{\alpha\beta} = \frac{\alpha^2 + \beta^2}{\alpha\beta} + \frac{\alpha^2 + \gamma^2}{\alpha\gamma} + \frac{\beta^2 + \gamma^2}{\beta\gamma}.$$

In particular, $\Sigma\alpha = \alpha + \beta + \gamma$, $\Sigma\alpha\beta$, and $\alpha\beta\gamma$ are called the three *elementary symmetric functions* of α, β, γ. In general,

$$\Sigma\alpha_1, \quad \Sigma\alpha_1\alpha_2, \quad \Sigma\alpha_1\alpha_2\alpha_3, \ldots, \quad \Sigma\alpha_1\alpha_2 \cdots \alpha_{n-1}, \quad \alpha_1\alpha_2 \ldots \alpha_n$$

are the elementary symmetric functions of $\alpha_1, \alpha_2, \ldots, \alpha_n$. In §20 they were written out more fully and proved to be equal to $-c_1$, c_2, $-c_3, \ldots, (-1)^n c_n$ if $\alpha_1, \ldots, \alpha_n$ are the roots of the equation

$$(1) \qquad x^n + c_1 x^{n-1} + c_2 x^{n-2} + \cdots + c_n = 0$$

whose leading coefficient is unity.

EXERCISES

If α, β, γ are the roots of $x^3 + px^2 + qx + r = 0$, so that $\Sigma\alpha = -p$, $\Sigma\alpha\beta = q$, and $\alpha\beta\gamma = -r$, prove that

1. $(\Sigma\alpha)^2 = \Sigma\alpha^2 + 2\Sigma\alpha\beta$, whence $\Sigma\alpha^2 = p^2 - 2q$.

2. $\Sigma\alpha \cdot \Sigma\alpha\beta = \Sigma\alpha^2\beta + 3\alpha\beta\gamma$, whence $\Sigma\alpha^2\beta = 3r - pq$.

3. $\Sigma\alpha^2\beta\gamma = pr$.

4. $\Sigma\alpha^2\beta^2 = (\Sigma\alpha\beta)^2 - 2\alpha\beta\gamma\Sigma\alpha = q^2 - 2pr$.

If α, β, γ, δ are the roots of $x^4 + px^3 + qx^2 + rx + s = 0$, prove that

5. $$\Sigma\frac{1}{\alpha} = \frac{-r}{s}, \qquad \Sigma\frac{1}{\alpha\beta} = \frac{q}{s}, \qquad \Sigma\frac{1}{\alpha^2} = \frac{r^2 - 2qs}{s^2}.$$
Hint: Compute the sum, sum of the products two at a time, and sum of the squares of the roots of the equation

$$1 + py + qy^2 + ry^3 + sy^4 = 0,$$

obtained by replacing x by $1/y$ in the given quartic equation.

6. $\Sigma\dfrac{\beta}{\alpha} = \Sigma\alpha \cdot \Sigma\dfrac{1}{\alpha} - 4 = \dfrac{pr}{s} - 4.$

7. $\Sigma\dfrac{\alpha^2 + \beta^2}{\alpha\beta} = \Sigma\dfrac{\beta}{\alpha}.$

8. $\Sigma\dfrac{\beta\gamma}{\alpha^2} = \Sigma\alpha\beta \cdot \Sigma\dfrac{1}{\alpha^2} - \Sigma\dfrac{\beta}{\alpha} = \dfrac{1}{s^2}(qr^2 - 2q^2s - prs + 4s^2).$

9. $\Sigma\dfrac{\gamma}{\alpha\beta} = \dfrac{3r - pq}{s}.$

104. Fundamental Theorem on Symmetric Functions. *Any polynomial symmetric in $x_1, \ldots, x_n$ is equal to an integral rational function, with integral coefficients, of the elementary symmetric functions*

(2) $\quad E_1 = \Sigma x_1, \qquad E_2 = \Sigma x_1 x_2, \qquad E_3 = \Sigma x_1 x_2 x_3, \ldots, \qquad E_n = x_1 x_2 \cdots x_n$

and the coefficients of the given polynomial. In particular, any symmetric polynomial with integral coefficients is equal to a polynomial in the elementary symmetric functions with integral coefficients.

For example, if $n = 2$,

$$rx_1^2 + rx_2^2 + sx_1 + sx_2 \equiv r(E_1^2 - 2E_2) + sE_1.$$

In case r and s are integers, the resulting polynomial in E_1 and E_2 has integral coefficients.

The theorem is most frequently used in the equivalent form:

Any polynomial symmetric in the roots of an equation,

$$x^n - E_1 x^{n-1} + E_2 x^{n-2} - \cdots + (-1)^n E_n = 0,$$

is equal to an integral rational function, with integral coefficients, of the coefficients of the equation and the coefficients of the polynomial.

It is this precise theorem that is required in all parts of modern algebra and the theory of numbers, where attention to the nature of the coefficients is vital, rather than the inadequate, oft-quoted, theorem that any symmetric function of the roots is expressible (rationally) in terms of the coefficients.

It suffices to prove the theorem for any homogeneous symmetric polynomial S, i.e., one expressible as a sum of terms

$$h = a x_1^{k_1} x_2^{k_2} \cdots x_n^{k_n}$$

of constant total degree $k = k_1 + k_2 + \cdots + k_n$ in the x's. Evidently we may assume that no two terms of S have the same set of exponents $k_1, \ldots, k_n$ (since such terms may be combined into a single one). We shall say that h is *higher* than the term $b x_1^{l_1} x_2^{l_2} \cdots x_n^{l_n}$ if $k_1 > l_1$, or if $k_1 = l_1$, $k_2 > l_2$, or if $k_1 = l_1$, $k_2 = l_2$, $k_3 > l_3, \ldots$, so that the first one of the differences $k_1 - l_1$, $k_2 - l_2$, $k_3 - l_3, \ldots$ which is not zero is positive.

We first prove that, if the above term h is the highest term of S, then

$$k_1 \geqq k_2 \geqq k_3 \cdots \geqq k_n.$$

For, if $k_1 < k_2$, the symmetric polynomial S would contain the term

$$a x_1^{k_2} x_2^{k_1} x_3^{k_3} \cdots x_n^{k_n},$$

which is higher than h. If $k_2 < k_3$, S would contain the term

$$a x_1^{k_1} x_2^{k_3} x_3^{k_2} \cdots x_n^{k_n},$$

which is higher than h, etc.

If the highest term in another homogeneous symmetric polynomial S' is

$$h' = a' x_1^{k_1'} x_2^{k_2'} \cdots x_n^{k_n'},$$

and that of S is h, then the highest term in their product SS' is

$$hh' = aa' x_1^{k_1 + k_1'} \cdots x_n^{k_n + k_n'}.$$

Indeed, suppose that SS' has a term, higher than hh',

(3)
$$cx_1^{l_1+l_1'} \cdots x_n^{l_n+l_n'},$$

which is either a product of terms

$$t = bx_1^{l_1} \cdots x_n^{l_n}, \qquad t' = b'x_1^{l_1'} \cdots x_n^{l_n'}$$

of S and S' respectively, or is a sum of such products. Since (3) is higher than hh', the first one of the differences

$$l_1 + l_1' - k_1 - k_1', \ldots, l_n + l_n' - k_n - k_n'$$

which is not zero is positive. But, either all of the differences $l_1 - k_1, \ldots, l_n - k_n$ are zero or the first one which is not zero is negative, since h is either identical with t or is higher than t. Likewise for the differences $l_1' - k_1', \ldots, l_n' - k_n'$. We therefore have a contradiction.

It follows at once that the highest term in a product of any number of homogeneous symmetric polynomials is the product of their highest terms. Now the highest terms in $E_1, E_2, E_3, \ldots, E_n$, given by (2), are

$$x_1, \qquad x_1 x_2, \qquad x_1 x_2 x_3, \qquad \ldots, \qquad x_1 x_2 \cdots x_n,$$

respectively. Hence the highest term in $E_1^{a_1} E_2^{a_2} \cdots E_n^{a_n}$ is

$$x_1^{a_1+a_2+\cdots+a_n} x_2^{a_2+\cdots+a_n} \cdots x_n^{a_n}.$$

Thus the highest term in

$$\sigma = aE_1^{k_1-k_2} E_2^{k_2-k_3} \cdots E_{n-1}^{k_{n-1}-k_n} E_n^{k_n}$$

is h. Hence $S_1 = S - \sigma$ is a homogeneous symmetric polynomial of the same total degree k as S and having a highest term h_1 not as high as h. As before, we form a product σ_1 of the E's whose highest term is this h_1. Then $S_2 = S_1 - \sigma_1$ is a homogeneous symmetric polynomial of total degree k and with a highest term h_2 not as high as h_1. We must finally reach a difference $S_t - \sigma_t$ which is identically zero. Indeed, there is only a finite number of products of powers of $x_1, \ldots, x_n$ of total degree k. Among these are the parts $h', h_1', h_2', \ldots$ of $h, h_1, h_2, \ldots$ with the coefficients suppressed. Since each h_i is not as high as h_{i-1}, the $h', h_1', h_2', \ldots$ are all distinct. Hence there is only a finite number of h_i. Since $S_t - \sigma_t \equiv 0$,

$$S = \sigma + S_1 = \sigma + \sigma_1 + S_2 = \cdots = \sigma + \sigma_1 + \sigma_2 + \cdots + \sigma_t.$$

Hence S is a polynomial in $E_1, E_2, \ldots, E_n$ and $a, b, \ldots$, with integral coefficients.

EXAMPLE 1.   If $S = \Sigma x_1^2 x_2^2 x_3$ and $n > 4$, we have

$$\sigma = E_2 E_3 = S + 3\Sigma x_1^2 x_2 x_3 x_4 + 10\Sigma x_1 x_2 x_3 x_4 x_5,$$
$$S_1 = S - \sigma = -3\Sigma x_1^2 x_2 x_3 x_4 - 10\Sigma x_1 x_2 x_3 x_4 x_5,$$
$$\sigma_1 = -3E_1 E_4 = -3(\Sigma x_1^2 x_2 x_3 x_4 + 5\Sigma x_1 x_2 x_3 x_4 x_5),$$
$$S_2 = S_1 - \sigma_1 = 5\Sigma x_1 x_2 x_3 x_4 x_5 = 5E_5,$$
$$S = \sigma + S_1 = \sigma + \sigma_1 + S_2 = E_2 E_3 - 3E_1 E_4 + 5E_5.$$

EXAMPLE 2.   If $S = \Sigma x_1^3 x_2 x_3$ and $n > 4$,

$$\sigma = E_1^2 E_3 = E_1(\Sigma x_1^2 x_2 x_3 + 4\Sigma x_1 x_2 x_3 x_4)$$
$$= \Sigma x_1^3 x_2 x_3 + 2\Sigma x_1^2 x_2^2 x_3 + 3\Sigma x_1^2 x_2 x_3 x_4$$
$$+ 4(\Sigma x_1^2 x_2 x_3 x_4 + 5\Sigma x_1 x_2 x_3 x_4 x_5),$$

$$S_1 = S - \sigma = -2\Sigma x_1^2 x_2^2 x_3 - 7\Sigma x_1^2 x_2 x_3 x_4 - 20\Sigma x_1 x_2 x_3 x_4 x_5.$$

Take $\sigma_1 = -2E_2 E_3$ and proceed as in Ex. 1.

EXAMPLE 3.   By examples 1 and 2, if $n > 4$,

$$a\Sigma x_1^2 x_2^2 x_3 + b\Sigma x_1^3 x_2 x_3 = bE_1^2 E_3 - (3a + b)E_1 E_4 + (a - 2b)E_2 E_3 + 5(a + b)E_5.$$

105. Rational Functions Symmetric in all but One of the Roots.

If P is a rational function of the roots of an equation $f(x) = 0$ of degree n and if P is symmetric in $n-1$ of the roots, then P is equal to a rational function, with integral coefficients, of the remaining root and the coefficients of $f(x)$ and P.

For example, $P = r\alpha_1 + \alpha_2^2 + \alpha_3^2 + \cdots + \alpha_n^2$ is symmetric in $\alpha_2, \ldots, \alpha_n$, and

$$P = r\alpha_1 + \Sigma\alpha_1^2 - \alpha_1^2 = c_1^2 - 2c_2 + r\alpha_1 - \alpha_1^2,$$

if $\alpha_1, \ldots, \alpha_n$ are the roots of equation (1).

Since[1] any symmetric rational function is the quotient of two symmetric polynomials, the above theorem will follow if proved for the case in which the words rational function are in both places replaced by polynomial.

---

[1] If N/D is symmetric in α_1, α_2, and the polynomials N and D have no common factor, while N becomes N' and D becomes D' when α_1, α_2 are interchanged, then $ND' \equiv DN'$. Thus N divides N' and both are of the same degree. Hence $N' = cN, D' = cD$, where c is a constant. By again interchanging α_1, α_2, we obtain N from N', whence $N = cN' = c^2 N$, $c^2 = 1$. If $c = -1$, we take $\alpha_1 = \alpha_2$ and see that $N = N' = -N$, $N = 0$, whence N has the factor $\alpha_1 - \alpha_2$. Similarly, D has the same factor, contrary to hypothesis. Hence $c = +1$ and N and D are each symmetric in α_1, α_2.

If α_1 is the remaining root, the polynomial P is symmetric in the roots $\alpha_2, \ldots, \alpha_n$ of $f(x)/(x - \alpha_1) = 0$, an equation of degree $n - 1$ whose coefficients are polynomials in $\alpha_1, c_1, \ldots, c_n$ with integral coefficients. Hence (§104), P is equal to a polynomial, with integral coefficients, in $\alpha_1, c_1, \ldots, c_n$ and the coefficients of P.

EXAMPLE.   If α, β, γ are the roots of $f(x) \equiv x^3 + px^2 + qx + r = 0$, find

$$\Sigma\frac{\alpha^2 + \beta^2}{\alpha + \beta} = \frac{\alpha^2 + \beta^2}{\alpha + \beta} + \frac{\alpha^2 + \gamma^2}{\alpha + \gamma} + \frac{\beta^2 + \gamma^2}{\beta + \gamma}.$$

Solution.   Since $\beta^2 + \gamma^2 = p^2 - 2q - \alpha^2$, $\beta + \gamma = -p - \alpha$,

$$\Sigma\frac{\alpha^2 + \beta^2}{\alpha + \beta} = \Sigma\frac{p^2 - 2q - \alpha^2}{-p - \alpha} = \Sigma\left(\alpha - p + \frac{2q}{\alpha + p}\right) = -p - 3p + 2q\Sigma\frac{1}{\alpha + p}.$$

But $\alpha + p, \beta + p, \gamma + p$ are the roots y_1, y_2, y_3 of the cubic equation obtained from $f(x) = 0$ by the substitution $x + p = y$, i.e., $x = y - p$. The resulting equation is

$$y^3 - 2py^2 + (p^2 + q)y + r - pq = 0.$$

Since we desire the sum of the reciprocals of y_1, y_2, y_3, we set $y = 1/z$ and find the sum of the roots z_1, z_2, z_3 of

$$1 - 2pz + (p^2 + q)z^2 + (r - pq)z^3 = 0.$$

Hence

$$\Sigma\frac{1}{\alpha + p} = \Sigma\frac{1}{y_1} = \Sigma z_1 = \frac{p^2 + q}{pq - r}, \qquad \Sigma\frac{\alpha^2 + \beta^2}{\alpha + \beta} = \frac{2q^2 - 2p^2 q + 4pr}{pq - r}.$$

EXERCISES

[In Exs. 1–12, α, β, γ are the roots of $f(x) = x^3 + px^2 + qx + r = 0$.]
Using $\beta\gamma + \alpha(\beta + \gamma) = q$, find

1. $\Sigma\dfrac{\beta\gamma + \alpha^2}{\beta + \gamma}$,

2. $\Sigma\dfrac{3\beta\gamma - 2\alpha^2}{\beta + \gamma - \alpha}$.

3. Why would the use of $\beta\gamma = -r/\alpha$ complicate Exs. 1, 2? Verify that

$$\beta\gamma = \frac{-r}{\alpha} = \frac{f(\alpha) - r}{\alpha} = \alpha^2 + p\alpha + q.$$

4. Why would you use $\beta\gamma = -r/\alpha$ in finding $\Sigma\dfrac{\beta^2 + \gamma^2}{\beta\gamma + c}$?

5. Find $\Sigma(\beta+\gamma)^2$.   **6.** Find $\Sigma(\alpha+\beta-\gamma)^3$.   **7.** Find $\Sigma\left(\dfrac{\beta-\gamma}{\beta+\gamma}\right)^2$.

8. Find a necessary and sufficient condition on the coefficients that the roots, in some order, shall be in harmonic progression. Hint: If $\dfrac{1}{\alpha}+\dfrac{1}{\gamma}=\dfrac{2}{\beta}$, then $\dfrac{-3r}{q}-\beta=0$, and conversely. Hence the condition is

$$\left(\frac{-3r}{q}-\alpha\right)\left(\frac{-3r}{q}-\beta\right)\left(\frac{-3r}{q}-\gamma\right)=f\left(\frac{-3r}{q}\right)=0.$$

9. Find the cubic equation with the roots $\beta\gamma-\dfrac{1}{\alpha}$, $\alpha\gamma-\dfrac{1}{\beta}$, $\alpha\beta-\dfrac{1}{\gamma}$. Hint: since these are $(-r-1)/\alpha$, etc., make the substitution $(-r-1)/x=y$.

Find the substitution which replaces the given cubic equation by one with the roots

10. $\alpha\beta+\alpha\gamma$, $\alpha\beta+\beta\gamma$, $\alpha\gamma+\beta\gamma$.

11. $\dfrac{2\alpha-1}{\beta+\gamma-\alpha}$, etc.   **12.** $\dfrac{\beta\gamma+3\alpha^2}{\beta+\gamma-2\alpha}$, etc.

If $\alpha,\beta,\gamma,\delta$ are the roots of $x^4+px^3+qx^2+rx+s=0$, find

13. $\Sigma\dfrac{\beta^2+\gamma^2+\delta^2}{\beta+\gamma+\delta}$.   **14.** $\Sigma\dfrac{\beta\gamma+\beta\delta+\gamma\delta}{\beta+\gamma+\delta-3}$.

15. Prove that if y_1, y_2, y_3 are the roots of $y^3+py+q=0$, the equation with the roots $z_1=(y_2-y_3)^2$, $z_2=(y_1-y_3)^2$, $z_3=(y_1-y_2)^2$ is

$$z^3+6pz^2+9p^2z+4p^3+27q^2=0.$$

Hints: since $z_1=\Sigma y_1^2-2y_2y_3-y_1^2=-2p+2q/y_1-y_1^2$, etc., we set $z=-2p+2q/y-y^2$. By the given equation, $y^2+p+q/y=0$. Thus the desired substitution is $z=-p+3q/y$, $y=3q/(z+p)$.

16. Hence find the discriminant of the reduced cubic equation.

17. If $x_1,\dots,x_n$ are the roots of $f(x)=0$, show that

$$\Sigma\frac{1}{x_1-c}=\frac{-f'(c)}{f(c)}.$$

Hint: $x_1-c=y_1,\dots,x_n-c=y_n$ are the roots of

$$f(c+y)=f(c)+yf'(c)+y^2(\ \ )+\cdots=0,$$

as shown by Taylor's theorem. Or we may employ (5) below for $x=c$.

106. Sums of Like Powers of the Roots. If $\alpha_1, \ldots, \alpha_n$ are the roots of

(1) $$f(x) \equiv x^n + c_1 x^{n-1} + c_2 x^{n-2} + \cdots + c_n = 0,$$

we write s_1 for $\Sigma \alpha_1$, s_2 for $\Sigma \alpha_1^2$, and, in general,

$$s_k = \Sigma \alpha_1^k = \alpha_1^k + \alpha_2^k + \cdots + \alpha_n^k.$$

The factored form of (1) is

(4) $$f(x) \equiv (x - \alpha_1)(x - \alpha_2) \cdots (x - \alpha_n).$$

The derivative $f'(x)$ of this product is found by multiplying the derivative (unity) of each factor by the product of the remaining factors and adding the results. Hence

$$f'(x) = (x - \alpha_2) \cdots (x - \alpha_n) + (x - \alpha_1)(x - \alpha_3) \cdots (x - \alpha_n) + \cdots,$$

(5) $$f'(x) \equiv \frac{f(x)}{x - \alpha_1} + \frac{f(x)}{x - \alpha_2} + \cdots + \frac{f(x)}{x - \alpha_n}.$$

If α is any root of (1), $f(\alpha) = 0$ and

$$\frac{f(x)}{x - \alpha} = \frac{f(x) - f(\alpha)}{x - \alpha} = \frac{x^n - \alpha^n}{x - \alpha} + c_1 \frac{x^{n-1} - \alpha^{n-1}}{x - \alpha} + \cdots + c_{n-1} \frac{x - \alpha}{x - \alpha}$$
$$= x^{n-1} + \alpha x^{n-2} + \alpha^2 x^{n-3} + \cdots + c_1(x^{n-2} + \alpha x^{n-3} + \cdots)$$
$$+ c_2(x^{n-3} + \cdots) + \cdots,$$

(6) $$\frac{f(x)}{x - \alpha} = x^{n-1} + (\alpha + c_1)x^{n-2} + (\alpha^2 + c_1\alpha + c_2)x^{n-3} + \cdots$$
$$+ (\alpha^k + c_1\alpha^{k-1} + c_2\alpha^{k-2} + \cdots + c_{k-1}\alpha + c_k)x^{n-k-1} + \cdots.$$

Taking α to be $\alpha_1, \ldots, \alpha_n$ in turn, adding the results, and applying (5), we obtain

$$f'(x) = nx^{n-1} + (s_1 + nc_1)x^{n-2} + (s_2 + c_1 s_1 + nc_2)x^{n-3} + \cdots$$
$$+ (s_k + c_1 s_{k-1} + c_2 s_{k-2} + \cdots + c_{k-1} s_1 + nc_k)x^{n-k-1} + \cdots.$$

The derivative of (1) is found at once by the rules of calculus (or by §56) to be

$$f'(x) = nx^{n-1} + (n-1)c_1 x^{n-2} + (n-2)c_2 x^{n-3} + \cdots + (n-k)c_k x^{n-k-1} + \cdots.$$

Since this expression is identical term by term with the preceding, we have

(7) $$s_1 + c_1 = 0, \qquad s_2 + c_1 s_1 + 2c_2 = 0, \ldots,$$
$$s_k + c_1 s_{k-1} + c_2 s_{k-2} + \cdots + c_{k-1} s_1 + kc_k = 0 \quad (k \leqq n - 1).$$

We may therefore find in turn $s_1, s_2, \ldots, s_{n-1}$:

(8) $s_1 = -c_1,$ $s_2 = c_1^2 - 2c_2,$ $s_3 = -c_1^3 + 3c_1 c_2 - 3c_3, \ldots .$

To find s_n, replace x in (1) by $\alpha_1, \ldots, \alpha_n$ in turn and add the resulting equations. We get

(9) $$s_n + c_1 s_{n-1} + c_2 s_{n-2} + \cdots + c_{n-1} s_1 + n c_n = 0.$$

We may combine (7) and (9) into

(10) $s_k + c_1 s_{k-1} + c_2 s_{k-2} + \cdots + c_{k-1} s_1 + k c_k = 0$ $(k = 1, 2, \ldots, n).$

This set of formulas (10) will be referred to as *Newton's identities*. The student should be able to write them down from memory and, when writing them, should always check the final one (9) by deriving it as above.

To derive a formula which shall enable us to compute the s_k for $k > n$, we multiply (1) by x^{k-n}, take $x = \alpha_1, \ldots, x = \alpha_n$ in turn, and add the resulting equations. We get

(11) $s_k + c_1 s_{k-1} + c_2 s_{k-2} + \cdots + c_n s_{k-n} = 0$ $(k > n).$

Instead of memorizing this formula, it is preferable to deduce it for the particular equation for which it is needed, thus avoiding errors of substitution as well as confusion with (10).

EXAMPLE.   Find s_k for $x^n - 1 = 0.$

Solution.   Comparing our equation with (1), we have $c_1 = 0, \ldots, c_{n-1} = 0,$ $c_n = -1$. Hence in (10) for $k < n$, each c is zero and $s_k = 0$. But, for $k = n$, (10) becomes $s_n - n = 0$. We may check the latter by substituting each root $\alpha_1, \ldots, \alpha_n$ in our given equation and adding. Finally, to find s_l when $l > n$, multiply our equation by x^{l-n}. In the resulting equation $x^l - x^{l-n} = 0$ we substitute each root, add, and obtain $s_l - s_{l-n} = 0$. Hence from s_l we obtain an equal s by subtracting n from l. After repeated subtractions, we reach a value k for which $1 \leq k \leq n$. Since $s_k = 0$ or n according as $k < n$ or $k = n$, it follows that $s_l = 0$ or n according as l is not or is divisible by n.

EXERCISES

1. For a cubic equation, $s_4 = c_1^4 - 4c_1^2 c_2 + 4c_1 c_3 + 2c_2^2.$

2. For an equation of degree ≥ 4, $s_4 = c_1^4 - 4c_1^2 c_2 + 4c_1 c_3 + 2c_2^2 - 4c_4.$

3. Find s_2, s_3, s_4, s_5 for $x^2 - px + q = 0.$

4. Find s_k for $x^5 - 3 = 0.$

5. Find s_2, s_3, s_6, s_7 for $x^5 - px + q = 0.$

107. Waring's Formula for s_k in Terms of the Coefficients. While we have learned how to find s_1, s_2, s_3,... in turn by Newton's identities, it is occasionally useful to know an explicit expression for s_k, where k has an arbitrary value. The formula in question is applied ordinarily only to a quadratic equation

$$x^2 + px + q = 0.$$

Accordingly we shall treat this case in detail. If its roots are α and β, then

$$x^2 + px + q \equiv (x - \alpha)(x - \beta).$$

Replace x by $1/y$ and multiply by y^2. We get

(12) $$1 + py + qy^2 \equiv (1 - \alpha y)(1 - \beta y).$$

Taking derivatives, we have

$$p + 2qy \equiv -\alpha(1 - \beta y) - \beta(1 - \alpha y).$$

Change of signs and division by the members of (12) gives

(13) $$\frac{-p - 2qy}{1 + py + qy^2} \equiv \frac{\alpha}{1 - \alpha y} + \frac{\beta}{1 - \beta y}.$$

The identity in Ex. 7, §14, with n changed to k, may be written in the form

(14) $$\frac{1}{1 - r} \equiv 1 + r + r^2 + \cdots + r^{k-1} + \frac{r^k}{1 - r}.$$

Take $r = \alpha y$ and multiply the resulting terms by α; thus

$$\frac{\alpha}{1 - \alpha y} = \alpha + \alpha^2 y + \cdots + \alpha^k y^{k-1} + \frac{\alpha^{k+1} y^k}{1 - \alpha y}.$$

Similarly,

$$\frac{\beta}{1 - \beta y} = \beta + \beta^2 y + \cdots + \beta^k y^{k-1} + \frac{\beta^{k+1} y^k}{1 - \beta y}.$$

To show that on adding, and writing s_k for $\alpha^k + \beta^k$, we obtain (15), we need the sum of the final fractions, which by (12) is

$$\frac{\phi y^k}{(1 - \alpha y)(1 - \beta y)} = \frac{\phi y^k}{1 + py + qy^2}, \qquad \phi \equiv \alpha^{k+1}(1 - \beta y) + \beta^{k+1}(1 - \alpha y).$$

Hence

(15) $$\frac{\alpha}{1 - \alpha y} + \frac{\beta}{1 - \beta y} = s_1 + s_2 y + \cdots + s_k y^{k-1} + \frac{\phi y^k}{1 + py + qy^2},$$

where the exact expression for ϕ is immaterial.

Next, we seek an expansion of the fraction in the left member of (13). Its denominator will be identical with that in (14) if we choose $r = -py - qy^2$. Evidently (14) may be written in the compact form

$$\frac{1}{1-r} \equiv \sum_{t=0}^{k-1} r^t + \frac{r^k}{1-r}.$$

Hence it becomes

$$\frac{1}{1+py+qy^2} = \sum_{t=0}^{k-1}(-1)^t(py+qy^2)^t + \frac{\psi y^k}{1+py+qy^2},$$

where $\psi = (-p-qy)^k$, although no use will be made of the particular form of the polynomial ψ. By the binomial theorem,

$$(py+qy^2)^t = \sum \frac{(g+h)!}{g!h!}(py)^g(qy^2)^h,$$

where the summation extends over all sets of integers g and h, each ≥ 0, for which $g+h = t$, while $g!$ denotes the product of $1, 2, \ldots, g$ if $g \geq 1$, but denotes unity if $g = 0$. Hence

(16) $\quad \dfrac{-p-2qy}{1+py+qy^2} = (p+2qy)\sum(-1)^{g+h+1}\dfrac{(g+h)!}{g!h!}p^g q^h y^{g+2h} + E,$

$$E \equiv \frac{(-p-2qy)\psi y^k}{1+py+qy^2},$$

where the summation extends over all sets of integers g and h, each ≥ 0, for which $g+h \leq k-1$.

Since the left members of (15) and (16) are identically equal by (13), their right members must be identical, so that the coefficients of y^{k-1} in them must be equal.[2] Hence the coefficient s_k of y^{k-1} in (15) is equal to the coefficient of y^{k-1} in (16), which is made up of two parts, corresponding to the two terms of the factor $p+2qy$. When we use the constant term p, we must employ from $\sum$ in (16) the terms in which the exponent of y is equal to $k-1$. But when we use the other term $2qy$, we must employ from $\sum$ the terms in which the exponent of y is equal to $k-2$, in order to obtain the combined exponent $k-1$ of y.

---

[2]In fact, the $(k-1)$th derivatives of the two right members are identical, and we obtain the indicated result by substituting $y = 0$ in these two derivatives and equating the results. Note that the final terms in both (15) and (16) have y as a factor of their $(k-1)$th derivatives.

Hence s_k is equal to the sum of the following two parts:

$$p\sum(-1)^{g+h+1}\frac{(g+h)!}{g!h!}p^g q^h \qquad (g+2h=k-1),$$

$$2q\sum(-1)^{g+h+1}\frac{(g+h)!}{g!h!}p^g q^h \qquad (g+2h=k-2).$$

In the upper sum, write i for $g+1$, and j for h. In the lower sum, write i for g, and j for $h+1$. Hence

$$s_k = \sum(-1)^{i+j}\frac{(i+j-1)!}{(i-1)!j!}p^i q^j + 2\sum(-1)^{i+j}\frac{(i+j-1)!}{i!(j-1)!}p^i q^j,$$

where now each summation extends over all sets of integers i and j, each ≥ 0, for which

$$(17) \qquad\qquad\qquad i+2j = k.$$

Finally, we may combine our two sums. Multiply the numerator and denominator of the first fraction by i, and those of the second fraction by j. Thus

$$(18) \qquad\qquad s_k = k\sum(-1)^{i+j}\frac{(i+j-1)!}{i!j!}p^i q^j,$$

since the present fraction occurred first multiplied by i and second multiplied by $2j$, and, by (17), the sum of these multipliers is equal to k. Our final result is (18), where the summation extends over all sets of integers i and j, each ≥ 0, satisfying (17).

If we replace i by its value $k-2j$, and change the sign of p, we obtain from (18) the result that *the sum of the kth powers of the roots of $x^2 - px + q = 0$ is equal to*

$$(19) \qquad s_k = k\sum_{j=0}^{K}(-1)^j\frac{(k-j-1)!}{(k-2j)!j!}p^{k-2j}q^j$$

$$= p^k - kp^{k-2}q + \frac{k(k-3)}{1\cdot 2}p^{k-4}q^2 - \frac{k(k-4)(k-5)}{1\cdot 2\cdot 3}p^{k-6}q^3 + \cdots,$$

where K is the largest integer not exceeding $k/2$.

The product of the roots is equal to q. Hence if x denotes one root, the second root is q/x. Thus $s_k = x^k + (q/x)^k$. Again, the sum of the roots is $x + q/x = p$. Regard q as given and p as unknown. Hence, if c is an arbitrary constant, the equation

$$(20) \qquad\qquad p^k - kqp^{k-2} + \frac{k(k-3)}{1\cdot 2}q^2 p^{k-4} - \cdots = c$$

is transformed by the substitution $p = x + q/x$ into

$$x^k + \left(\frac{q}{x}\right)^k = c.$$

Hence equation (20) may be solved for p by radicals by the method employed in §43 for a cubic equation.

The above proof applies[3] without essential change to any equation $x^n + c_1 x^{n-1} + \cdots + c_n = 0$ and leads to the following formula for the sum of the kth powers of its roots:

$$(21) \qquad s_k = k \sum (-1)^{r_1 + \cdots + r_n} \frac{(r_1 + \cdots + r_n - 1)!}{r_1! \cdots r_n!} c_1^{r_1} \cdots c_n^{r_n},$$

where the summation extends over all sets of integers $r_1, \ldots, r_n$, each ≥ 0, for which $r_1 + 2r_2 + 3r_3 + \cdots + nr_n = k$. This result (21) is known as *Waring's formula* and was published by him in 1762.

EXAMPLE.   Let $n = 3$, $k = 4$. Then $r_1 + 2r_2 + 3r_3 = 4$ and

$$(r_1, r_2, r_3) = (4, 0, 0), \qquad (2, 1, 0), \qquad (1, 0, 1), \qquad (0, 2, 0),$$

$$s_4 = 4\left(\frac{3!}{4!}c_1^4 - \frac{2!}{2!1!}c_1^2 c_2 + \frac{1!}{1!1!}c_1 c_3 + \frac{1!}{2!}c_2^2\right)$$

$$= c_1^4 - 4c_1^2 c_2 + 4c_1 c_3 + 2c_2^2.$$

EXERCISES

1. For the quadratic $x^2 - px + q = 0$ write out the expressions for s_2, s_3, s_4, s_5 given by (19), and compare with those obtained from Newton's identities (Ex. 3, §106).

2. Find s_4 for a quartic equation by Waring's formula.

3. For $k = 5$, (20) becomes De Moivre's quintic $p^5 - 5qp^3 + 5q^2 p = c$. Solve it by radicals for p.

4. Solve (20) by radicals when $k = 7$.

---

[3]See the author's *Elementary Theory of Equations*, pp. 72–74, where there is given also a shorter proof by means of infinite series.

108. Σ-functions Expressed in Terms of the Functions s_k. Since we have learned two methods of expressing the s_k in terms of the coefficients, it is desirable to learn how to express any Σ-polynomial (and hence any symmetric function) in terms of the s_k.

By performing the indicated multiplication, we find that

$$s_a s_b \equiv \Sigma \alpha_1^a \cdot \Sigma \alpha_1^b = \Sigma \alpha_1^{a+b} + m\Sigma \alpha_1^a \alpha_2^b,$$

where $m = 1$ if $a \neq b$, $m = 2$ if $a = b$. Transposing the first term, which is equal to s_{a+b}, and dividing by m, we obtain

$$(22) \qquad \Sigma \alpha_1^a \alpha_2^b = \frac{1}{m}(s_a s_b - s_{a+b}).$$

In order to compute $\Sigma \alpha_1^4 \alpha_2^3 \alpha_3^2$ in terms of the s_k, we form the product

$$\Sigma \alpha_1^4 \cdot \Sigma \alpha_1^3 \alpha_2^2 = \Sigma \alpha_1^7 \alpha_2^2 + \Sigma \alpha_1^6 \alpha_2^3 + \Sigma \alpha_1^4 \alpha_2^3 \alpha_3^2.$$

Making three applications of (22), we get

$$s_4(s_3 s_2 - s_5) = (s_7 s_2 - s_9) + (s_6 s_3 - s_9) + \Sigma \alpha_1^4 \alpha_2^3 \alpha_3^2.$$

Hence

$$\Sigma \alpha_1^4 \alpha_2^3 \alpha_3^2 = s_2 s_3 s_4 - s_2 s_7 - s_3 s_6 - s_4 s_5 + 2s_9.$$

EXERCISES

For a quartic equation, express in terms of the s_k and ultimately in terms of the coefficients $c_1, \ldots, c_4$:

1. $\Sigma \alpha_1^2 \alpha_2^2.$ **2.** $\Sigma \alpha_1^3 \alpha_2.$ **3.** $\Sigma \alpha_1^2 \alpha_2 \alpha_3.$ **4.** $\Sigma \alpha_1^2 \alpha_2^2 \alpha_3^2.$

5. If $a \geq b > c > 0$, prove that

$$\Sigma \alpha_1^a \alpha_2^b \alpha_3^c = \frac{1}{m}(s_a s_b s_c - s_a s_{b+c} - s_b s_{a+c} - s_c s_{a+b} + 2s_{a+b+c}),$$

where $m = 1$ if $a > b$, $m = 2$ if $a = b$.

6. $\Sigma \alpha_1^a \alpha_2^b \alpha_3^b = \frac{1}{2}(s_a s_b^2 - s_a s_{2b} - 2s_b s_{a+b} + 2s_{a+2b})$, $a > b > 0.$

7. $\Sigma \alpha_1^a \alpha_2^a \alpha_3^a = \frac{1}{6}(s_a^3 - 3s_a s_{2a} + 2s_{3a})$, $a > 0.$

109. Computation of Symmetric Functions.   The method last explained is practicable when a term of the Σ-function involves only a few distinct roots, the largeness of the exponents not introducing a difficulty in the initial work of expressing the Σ-function in terms of the s_k.

But when a term of the Σ-function involves a large number of roots with small exponents, we resort to a method suggested by §104, which tells us which auxiliary simpler symmetric functions should be multiplied together to produce our Σ-function along with simpler ones.

For example, to find $\Sigma x_1^2 x_2 x_3 x_4$, when $n > 4$, we employ

$$E_1 E_4 \equiv \Sigma x_1 \cdot \Sigma x_1 x_2 x_3 x_4 = \Sigma x_1^2 x_2 x_3 x_4 + 5 \Sigma x_1 x_2 x_3 x_4 x_5,$$

$$\Sigma x_1^2 x_2 x_3 x_4 = E_1 E_4 - 5 E_5.$$

To find $\Sigma x_1^2 x_2^2 x_3^2 x_4$, employ $E_3 E_4 = \Sigma x_1 x_2 x_3 \cdot \Sigma x_1 x_2 x_3 x_4$.

When many such products of Σ-functions are to be computed, it will save time in the long run to learn and apply the "method of leaders" explained in the author's *Elementary Theory of Equations*, pp. 64–65.

MISCELLANEOUS EXERCISES

Express in terms of the coefficients $c_1, \ldots, c_n$:

1. $\Sigma \alpha_1^2 \alpha_2 \alpha_3$.   **2.** $\Sigma \alpha_1^2 \alpha_2^2 \alpha_3$.   **3.** $\Sigma \alpha_1^2 \alpha_2^2 \alpha_3 \alpha_4$.   **4.** $\Sigma \alpha_1^2 \alpha_2^2 \alpha_3^2$.

If α, β, γ are the roots of $x^3 + px^2 + qx + r = 0$, find a cubic equation with the roots

5. α^2, β^2, γ^2.   **6.** $\alpha\beta$, $\alpha\gamma$, $\beta\gamma$.   **7.** $\dfrac{2}{\alpha}, \dfrac{2}{\beta}, \dfrac{2}{\gamma}$.

8. $\alpha^2 + \beta^2$, $\alpha^2 + \gamma^2$, $\beta^2 + \gamma^2$.   **9.** $\alpha^2 + \alpha\beta + \beta^2$, etc.

If α, β, γ, δ are the roots of $x^4 + px^3 + qx^2 + rx + s = 0$, find

10. $\Sigma \dfrac{\beta}{\alpha} = \Sigma \dfrac{\beta + \gamma + \delta}{\alpha} = \Sigma \dfrac{-p - \alpha}{\alpha} = -4 - p \Sigma \dfrac{1}{\alpha}$.

11. $\Sigma \dfrac{\beta}{\alpha^2}$. Use $\Sigma \dfrac{1}{\alpha} \cdot \Sigma \dfrac{\beta}{\alpha} = \Sigma \dfrac{\beta}{\alpha^2} + 3 \Sigma \dfrac{1}{\alpha} + 2 \Sigma \dfrac{\gamma}{\alpha\beta}$.

12. Express $\Sigma \alpha_1^a \alpha_2^b \alpha_3^c \alpha_4^d$ in terms of the s_k when (i) $a > b > c > d > 0$, and (ii) when $a = b = c = d$.

13. By solving the first k of Newton's identities (10) as a system of linear equations, find an expression in the form of a determinant (i) for s_k in terms of $c_1, \ldots, c_k$, and (ii) for c_k in terms of $s_1, \ldots, s_k$.

14. One set of n numbers is a mere rearrangement of another set if $s_1, \ldots, s_n$ have the same values for each set.

CHAPTER X

110. Elimination. If the two equations

$$ax + b = 0, \qquad cx + d = 0 \qquad\qquad (a \neq 0, \ c \neq 0)$$

are simultaneous, i.e., if x has the same value in each, then

$$x = -\frac{b}{a} = -\frac{d}{c}, \qquad R \equiv ad - bc = 0,$$

and conversely. Hence a necessary and sufficient condition that the equations have a common root is $R = 0$. We call R the *resultant* (or *eliminant*) of the two equations.

The result of eliminating x between the two equations might equally well have been written in the form $bc - ad = 0$. But the arbitrary selection of R as the resultant, rather than the product of R by some constant, as -1, is a matter of more importance than is apparent at first sight. For, we seek a *definite* function of the coefficients a, b, c, d of the *functions* $ax + b$, $cx + d$, and not merely a property $R = 0$ or $R \neq 0$ of the corresponding *equations*. Accordingly, we shall lay down the definition in §111, which, as the reader may verify, leads to R in our present example.

Methods of elimination which seem plausible often yield not R itself, but the product of R by an extraneous function of the coefficients. This point (illustrated in §114) indicates that the subject demands a more careful treatment than is often given.

111. Resultant of Two Polynomials in x. Let

(1)
$$\begin{cases} f(x) = a_0 x^m + a_1 x^{m-1} + \cdots + a_m & (a_0 \neq 0), \\ g(x) = b_0 x^n + b_1 x^{n-1} + \cdots + b_n & (b_0 \neq 0) \end{cases}$$

be two polynomials of degrees m and n. Let $\alpha_1, \ldots, \alpha_m$ be the roots of $f(x) = 0$. Since α_1 is a root of $g(x) = 0$ only when $g(\alpha_1) = 0$, the two equations have a root in common if and only if the product

$$g(\alpha_1)g(\alpha_2)\cdots g(\alpha_m)$$

is zero. This symmetric function of the roots of $f(x) = 0$ is of degree n in any one root and hence is expressible as a polynomial of degree n in the elementary symmetric functions (§104), which are equal to $-a_1/a_0$, a_2/a_0, ... To be rid of the denominators a_0, it therefore suffices to multiply our polynomial by a_0^n. We therefore define

$$(2) \qquad R(f,g) = a_0^n g(\alpha_1) g(\alpha_2) \cdots g(\alpha_m)$$

to be the *resultant* of f and g. It equals an integral rational function of $a_0, \ldots, a_m$, $b_0, \ldots, b_n$ with integral coefficients.

EXERCISES

1. If $m = 1$, $n = 2$, $R(f,g) = b_0 a_1^2 - b_1 a_0 a_1 + b_2 a_0^2$.

2. If $m = 2$, $n = 1$, $R(f,g) = a_0(b_0 \alpha_1 + b_1)(b_0 \alpha_2 + b_1) = a_0 b_1^2 - a_1 b_0 b_1 + a_2 b_0^2$, since

$$a_0(\alpha_1 + \alpha_2) = -a_1, \qquad a_0 \alpha_1 \alpha_2 = a_2.$$

3. If $\beta_1, \ldots, \beta_n$ are the roots of $g(x) = 0$, so that

$$g(\alpha_i) = b_0(\alpha_i - \beta_1)(\alpha_i - \beta_2) \cdots (\alpha_i - \beta_n),$$

then

$$\begin{aligned}
R(f,g) = a_0^n b_0^m (\alpha_1 - \beta_1)(\alpha_1 - \beta_2) \cdots (\alpha_1 - \beta_n) \\
\cdot (\alpha_2 - \beta_1)(\alpha_2 - \beta_2) \cdots (\alpha_2 - \beta_n) \\
\cdots\cdots\cdots\cdots\cdots\cdots\cdots\cdots \\
\cdot (\alpha_m - \beta_1)(\alpha_m - \beta_2) \cdots (\alpha_m - \beta_n).
\end{aligned}$$

Multiplying together the differences in each column, we see that

$$R(f,g) = (-1)^{mn} b_0^m f(\beta_1) f(\beta_2) \cdots f(\beta_n) = (-1)^{mn} R(g,f).$$

4. If $m = 2$, $n = 1$, $R(g,f) = b_0^2 f(-b_1/b_0) = a_0 b_1^2 - a_1 b_0 b_1 + a_2 b_0^2$, which is equal to $R(f,g)$ by Ex. 2. This illustrates the final result in Ex. 3.

5. If $m = n = 2$,

$$\begin{aligned}
R(f,g) &= a_0^2 b_0^2 \alpha_1^2 \alpha_2^2 + a_0^2 b_0 b_1 \alpha_1 \alpha_2 (\alpha_1 + \alpha_2) \\
&\quad + a_0^2 b_0 b_2 (\alpha_1^2 + \alpha_2^2) + a_0^2 b_1^2 \alpha_1 \alpha_2 + a_0^2 b_1 b_2 (\alpha_1 + \alpha_2) + a_0^2 b_2^2 \\
&= b_0^2 a_2^2 - b_0 b_1 a_1 a_2 + b_0 b_2 (a_1^2 - 2a_0 a_2) + b_1^2 a_0 a_2 - b_1 b_2 a_0 a_1 + a_0^2 b_2^2.
\end{aligned}$$

This equals $R(g,f)$, since it is unaltered when the a's and b's are interchanged.

6. Prove by (2) that R is homogeneous and of total degree m in $b_0, \ldots, b_n$; and by Ex. 3, that R is homogeneous and of total degree n in $a_0, \ldots, a_m$. Show that R has the terms $a_0^n b_n^m$ and $(-1)^{mn} b_0^m a_m^n$.

7. $R(f, g_1 g_2) = R(f, g_1) \cdot R(f, g_2)$.

8. $R(f, x^n) = (-1)^{mn} R(x^n, f) = (-1)^{mn} a_m^n$.

112. Sylvester's Dialytic Method of Elimination.[1] Let the equations

$$f(x) \equiv a_0 x^3 + a_1 x^2 + a_2 x + a_3 = 0, \qquad g(x) \equiv b_0 x^2 + b_1 x + b_2 = 0$$

have a common root x. Multiply the first equation by x and the second by x^2 and x in turn. We now have five equations

$$a_0 x^4 + a_1 x^3 + a_2 x^2 + a_3 x \qquad\quad = 0,$$
$$a_0 x^3 + a_1 x^2 + a_2 x + a_3 = 0,$$
$$b_0 x^4 + b_1 x^3 + b_2 x^2 \qquad\qquad = 0,$$
$$b_0 x^3 + b_1 x^2 + b_2 x \qquad\quad = 0,$$
$$b_0 x^2 + b_1 x + b_2 = 0,$$

which are linear and homogeneous in x^4, x^3, x^2, x, 1. Hence (§97)

(3)
$$F = \begin{vmatrix} a_0 & a_1 & a_2 & a_3 & 0 \\ 0 & a_0 & a_1 & a_2 & a_3 \\ b_0 & b_1 & b_2 & 0 & 0 \\ 0 & b_0 & b_1 & b_2 & 0 \\ 0 & 0 & b_0 & b_1 & b_2 \end{vmatrix}$$

must be zero. Next, if $F = 0$, there exist (§97) values which, when substituted for x^4, x^3, x^2, x and 1, satisfy the five equations. But why is the value for x^4 the fourth power of the value for x, that for x^3 the cube of the value for x, etc.? Since the direct verification of these facts would be very laborious, we resort to a device to show that, conversely, if $F = 0$ the two given equations have a root in common.

In (3) replace a_3 by $a_3 - z$ and consider the equation

(4)
$$\begin{vmatrix} a_0 & a_1 & a_2 & a_3 - z & 0 \\ 0 & a_0 & a_1 & a_2 & a_3 - z \\ b_0 & b_1 & b_2 & 0 & 0 \\ 0 & b_0 & b_1 & b_2 & 0 \\ 0 & 0 & b_0 & b_1 & b_2 \end{vmatrix} = 0.$$

To prove that it has the roots $f(\beta_1)$ and $f(\beta_2)$, where β_1 and β_2 are the roots of $g(x) = 0$, we take $z = f(\beta_i)$ and prove that the determinant is then equal to zero. For, if we add to the last column the products of the elements of the first four columns by β_i^4, β_i^3, β_i^2, β_i, respectively, we find that all of the elements of the new last column are zero.

---

[1] Given without proof by Sylvester, *Philosophical Magazine*, 1840, p. 132.

Since (4) reduces to (3) for $z = 0$, it is of the form

$$b_0^3 z^2 + kz + F = 0,$$

in which the value of k is immaterial. By considering the product of the roots of this quadratic equation, we see that

$$F = b_0^3 f(\beta_1) f(\beta_2).$$

Hence the Sylvester determinant F is the resultant $R(g, f)$ and hence is the resultant $R(f, g)$, since mn is here even (Ex. 3, §111).

In general, if the equations are

$$f(x) \equiv a_0 x^m + \cdots + a_m = 0, \qquad g(x) \equiv b_0 x^n + \cdots + b_n = 0,$$

we multiply the first equation by x^{n-1}, $x^{n-2}, \ldots, x$, 1, in turn, and the second by x^{m-1}, $x^{m-2}, \ldots, x$, 1, in turn. We obtain $n + m$ equations which are linear and homogeneous in the $m + n$ quantities $x^{m+n-1}, \ldots, x$, 1. Hence the determinant

$$(5) \quad F = \begin{vmatrix} a_0 & a_1 & a_2 & \ldots\ldots & a_m & 0 & \ldots\ldots\ldots & 0 \\ 0 & a_0 & a_1 & a_2 & \ldots\ldots & a_m & 0 & \ldots\ldots & 0 \\ 0 & 0 & a_0 & a_1 & a_2 & \ldots\ldots & a_m & 0 & \ldots & 0 \\ \ldots\ldots\ldots\ldots\ldots\ldots\ldots\ldots\ldots\ldots\ldots\ldots\ldots \\ 0 & \ldots\ldots & 0 & a_0 & a_1 & a_2 & \ldots\ldots\ldots & a_m \\ b_0 & b_1 & \ldots\ldots\ldots & b_n & 0 & \ldots\ldots\ldots & 0 \\ 0 & b_0 & b_1 & \ldots\ldots\ldots\ldots & b_n & \ldots & 0 \\ \ldots\ldots\ldots\ldots\ldots\ldots\ldots\ldots\ldots\ldots\ldots\ldots\ldots \\ 0 & \ldots & 0 & b_0 & b_1 & \ldots\ldots\ldots\ldots & b_n \end{vmatrix} \begin{matrix} \\ \left. \right\} n \text{ rows} \\ \\ \\ \left. \right\} m \text{ rows} \end{matrix}$$

is zero. It may be shown to be equal to the resultant $R(f, g)$, whether mn is even or odd, by the method employed in the above case $m = 3$, $n = 2$.

We may also prove as follows that if $F = 0$ the equations $f = 0$ and $g = 0$ have a common root. Since F was obtained as the determinant of the coefficients of

$$x^{n-1} f, \ldots, xf, f, \qquad x^{m-1} g, \ldots, xg, g,$$

$F = 0$ implies, by §96, Lemma 2, the existence of a linear relation

$$B_0 x^{n-1} f + \cdots + B_{n-2} xf + B_{n-1} f + A_0 x^{m-1} g + \cdots + A_{m-2} xg + A_{m-1} g \equiv 0,$$

identically in x, with constant coefficients $B_0, \ldots, A_{m-1}$ not all zero. In other words, $\beta f + \alpha g \equiv 0$, where

$$(6) \quad \alpha \equiv A_0 x^{m-1} + \cdots + A_{m-2} x + A_{m-1}, \qquad \beta \equiv B_0 x^{n-1} + \cdots + B_{n-2} x + B_{n-1}.$$

Neither α nor β is identically zero. For, if $\alpha \equiv 0$, for example, then $\beta f \equiv 0$ and $\beta \equiv 0$, whereas the A_i and B_i are not all zero.

Consider the factored forms of f, g, α, β. Suppose that f and g have no common linear factor. The highest power of each linear factor occurring in f divides $\alpha g \equiv -\beta f$ and hence divides α. Thus f divides α, whereas f is of higher degree than α. Hence our assumption that $f = 0$ and $g = 0$ have no common root has led to a contradiction.

A similar idea is involved in the method of elimination due to Euler (1707–1783). If $f = 0$ and $g = 0$ have a common root c, then $f \equiv (x - c)\alpha$, $-g \equiv (x - c)\beta$, identically in x, where α and β are polynomials in x of degrees $m - 1$ and $n - 1$, respectively. Give them the notations (6). In the identity $\beta f + \alpha g \equiv 0$, the coefficient of each power of x is zero. Hence

$$
\begin{aligned}
a_0 B_0 &\qquad + b_0 A_0 &&= 0 \\
a_1 B_0 + a_0 B_1 &\qquad + b_1 A_0 + b_0 A_1 &&= 0 \\
&\;\cdots\cdots\cdots\cdots\cdots\cdots\cdots\cdots \\
a_m B_{n-2} + a_{m-1} B_{n-1} &\qquad + b_n A_{m-2} + b_{n-1} A_{m-1} &&= 0 \\
a_m B_{n-1} &\qquad + b_n A_{m-1} &&= 0.
\end{aligned}
$$

Since these $m + n$ linear homogeneous equations in the unknowns $B_0, \ldots, B_{n-1}$, $A_0, \ldots, A_{m-1}$ have a set of solutions not all zero, the determinant of the coefficients is zero. By interchanging the rows and columns, we obtain the determinant (5).

EXERCISES

1. For $m = n = 2$, show that the resultant is

$$
R = \begin{vmatrix}
a_0 & a_1 & a_2 & 0 \\
0 & a_0 & a_1 & a_2 \\
b_0 & b_1 & b_2 & 0 \\
0 & b_0 & b_1 & b_2
\end{vmatrix}
$$

Interchange the second and third rows, apply Laplace's development, and prove that

$$
R = (a_0 b_2)^2 - (a_0 b_1)(a_1 b_2),
$$

where $(a_0 b_2)$ denotes $a_0 b_2 - a_2 b_0$, etc.

2. For $m = n = 3$, write down the resultant R and, by interchanges of rows, derive the second determinant in

$$
R = \begin{vmatrix}
a_0 & a_1 & a_2 & a_3 & 0 & 0 \\
0 & a_0 & a_1 & a_2 & a_3 & 0 \\
0 & 0 & a_0 & a_1 & a_2 & a_3 \\
b_0 & b_1 & b_2 & b_3 & 0 & 0 \\
0 & b_0 & b_1 & b_2 & b_3 & 0 \\
0 & 0 & b_0 & b_1 & b_2 & b_3
\end{vmatrix}
= - \begin{vmatrix}
a_0 & a_1 & a_2 & a_3 & 0 & 0 \\
b_0 & b_1 & b_2 & b_3 & 0 & 0 \\
0 & a_0 & a_1 & a_2 & a_3 & 0 \\
0 & b_0 & b_1 & b_2 & b_3 & 0 \\
0 & 0 & a_0 & a_1 & a_2 & a_3 \\
0 & 0 & b_0 & b_1 & b_2 & b_3
\end{vmatrix}
$$

To the second determinant apply Laplace's development, selecting minors from the first two rows, and to the complementary minors apply a similar development. This may be done by inspection and the following value of $-R$ will be obtained:

$$(a_0b_1)\{(a_1b_2)(a_2b_3) - (a_1b_3)^2 + (a_2b_3)(a_0b_3)\}$$
$$-(a_0b_2)\{(a_0b_2)(a_2b_3) - (a_0b_3)(a_1b_3)\}$$
$$+(a_0b_3)\{(a_0b_1)(a_2b_3) - (a_0b_3)^2\}.$$

The third term of the first line and the first term of the last line are alike. Hence, changing the signs,

$$R = (a_0b_3)^3 - 2(a_0b_1)(a_0b_3)(a_2b_3) - (a_0b_2)(a_0b_3)(a_1b_3)$$
$$+ (a_0b_2)^2(a_2b_3) + (a_0b_1)(a_1b_3)^2 - (a_0b_1)(a_1b_2)(a_2b_3).$$

Other methods of simplifying Sylvester's determinant (5) are given in §113.

113. Bézout's Method of Elimination.   When the two equations are of the same degree, the method published by Bézout in 1764 will be clear from the example

$$f \equiv a_0x^3 + a_1x^2 + a_2x + a_3 = 0, \qquad g \equiv b_0x^3 + b_1x^2 + b_2x + b_3 = 0.$$

Then

$$a_0g - b_0f,$$
$$(7) \qquad (a_0x + a_1)g - (b_0x + b_1)f,$$
$$(a_0x^2 + a_1x + a_2)g - (b_0x^2 + b_1x + b_2)f$$

are equal respectively to

$$(8) \qquad \begin{aligned} (a_0b_1)x^2 & \quad\quad + (a_0b_2)\,x + (a_0b_3) = 0, \\ (a_0b_2)x^2 & + \{(a_0b_3) + (a_1b_2)\}x + (a_1b_3) = 0, \\ (a_0b_3)x^2 & \quad\quad + (a_1b_3)\,x + (a_2b_3) = 0, \end{aligned}$$

where $(a_0b_1) = a_0b_1 - a_1b_0$, etc. The determinant of the coefficients is the negative of the resultant $R(f, g)$. Indeed, the negative of the determinant is easily verified to have the expansion given at the end of Ex. 2 just above.

To give a more instructive proof of the last fact, note that, by (7), equations (8) are linear combinations of

$$x^2f = 0, \qquad xf = 0, \qquad f = 0, \qquad x^2g = 0, \qquad xg = 0, \qquad g = 0,$$

the latter being the equations used in Sylvester's method of elimination. The determinant of the coefficients in these six equations is the first determinant R in Ex. 2

just above. The operations carried out to obtain equations (8) are seen to correspond step for step to the following operations on determinants. To the products of the elements of the fourth row by a_0 add the products of the elements of the 1st, 2nd, 3rd, 5th, 6th rows by $-b_0$, $-b_1$, $-b_2$, a_1, a_2 respectively [corresponding to the formation of the third function (7)]. To the products of the elements of the fifth row by a_0 add the products of the elements of the 2nd, 3rd, 6th rows by $-b_0$, $-b_1$, a_1 respectively [corresponding to the second function (7)]. Finally, to the products of the elements of the sixth row by a_0 add the products of the elements of the third row by $-b_0$ [corresponding to $a_0 g - b_0 f$]. Hence

$$a_0^3 R = \begin{vmatrix} a_0 & a_1 & a_2 & a_3 & 0 & 0 \\ 0 & a_0 & a_1 & a_2 & a_3 & 0 \\ 0 & 0 & a_0 & a_1 & a_2 & a_3 \\ 0 & 0 & 0 & (a_0 b_3) & (a_1 b_3) & (a_2 b_3) \\ 0 & 0 & 0 & (a_0 b_2) & (a_0 b_3) + (a_1 b_2) & (a_1 b_3) \\ 0 & 0 & 0 & (a_0 b_1) & (a_0 b_2) & (a_0 b_3) \end{vmatrix},$$

so that R is equal to the 3-rowed minor enclosed by the dots. The method of Bézout therefore suggests a definite process for the reduction of Sylvester's determinant of order $2n$ (when $m = n$) to one of order n.

Next, for equations of different degrees, consider the example

$$f \equiv a_0 x^4 + a_1 x^3 + a_2 x^2 + a_3 x + a_4, \qquad g \equiv b_0 x^2 + b_1 x + b_2.$$

Then

$$a_0 x^2 g - b_0 f, \qquad (a_0 x + a_1) x^2 g - (b_0 x + b_1) f$$

are equal respectively to

$$(a_0 b_1) x^3 + (a_0 b_2) x^2 - a_3 b_0 x - a_4 b_0,$$
$$(a_0 b_2) x^3 + \{(a_1 b_2) - a_3 b_0\} x^2 - \{a_3 b_1 + a_4 b_0\} x - a_4 b_1.$$

The determinant of the coefficients of x^3, x^2, x, 1 in these two functions and xg, g, after the first and second rows are interchanged, is the determinant of order 4 enclosed by dots in the second determinant below. Hence it is the resultant $R(f, g)$.

As in the former example, we shall indicate the corresponding operations on Sylvester's determinant

$$R = \begin{vmatrix} a_0 & a_1 & a_2 & a_3 & a_4 & 0 \\ 0 & a_0 & a_1 & a_2 & a_3 & a_4 \\ b_0 & b_1 & b_2 & 0 & 0 & 0 \\ 0 & b_0 & b_1 & b_2 & 0 & 0 \\ 0 & 0 & b_0 & b_1 & b_2 & 0 \\ 0 & 0 & 0 & b_0 & b_1 & b_2 \end{vmatrix}$$

Multiply the elements of the third and fourth rows by a_0. In the resulting determinant $a_0^2 R$, add to the elements of the third row the products of the elements of the first, second and fourth rows by $-b_0$, $-b_1$, a_1/a_0 respectively. Add to the elements of the fourth row the products of those of the second by $-b_0$. We get

$$a_0^2 R = \begin{vmatrix} a_0 & a_1 & a_2 & a_3 & a_4 & 0 \\ 0 & a_0 & a_1 & a_2 & a_3 & a_4 \\ 0 & 0 & (a_0b_2) & (a_1b_2) - a_3b_0 & -a_3b_1 - a_4b_0 & -a_4b_1 \\ 0 & 0 & (a_0b_1) & (a_0b_2) & -a_3b_0 & -a_4b_0 \\ 0 & 0 & b_0 & b_1 & b_2 & 0 \\ 0 & 0 & 0 & b_0 & b_1 & b_2 \end{vmatrix}$$

Hence R is equal to the minor enclosed by dots.

EXERCISES

1. For $m = 3$, $n = 2$, apply to Sylvester's determinant R exactly the same operations as used in the last case in §113 and obtain

$$R = \begin{vmatrix} (a_0b_2) & (a_1b_2) - a_3b_0 & -a_3b_1 \\ (a_0b_1) & (a_0b_2) & -a_3b_0 \\ b_0 & b_1 & b_2 \end{vmatrix}.$$

2. For $m = n = 4$, reduce Sylvester's R (as in the first case in §113) to

$$\begin{vmatrix} (a_0b_1) & (a_0b_2) & (a_0b_3) & (a_0b_4) \\ (a_0b_2) & (a_0b_3) + (a_1b_2) & (a_0b_4) + (a_1b_3) & (a_1b_4) \\ (a_0b_3) & (a_0b_4) + (a_1b_3) & (a_1b_4) + (a_2b_3) & (a_2b_4) \\ (a_0b_4) & (a_1b_4) & (a_2b_4) & (a_3b_4) \end{vmatrix}.$$

114. General Theorem on Elimination. *If any method of eliminating x between two equations in x leads to a relation $F = 0$, where F is a polynomial in the coefficients, then F has as a factor the true resultant of the equations.*

Some of the preceding proofs become simpler if this theorem is applied. For example, determinant (3) is divisible by the resultant R. Since the diagonal term of (3) is a term $a_0^2 b_2^3$ of R (Ex. 6, §111), F is identical with R.

The preceding general theorem is proved in the author's *Elementary Theory of Equations*, pp. 152–4. We shall here merely verify the theorem in an instructive special case. Let

$$f \equiv a_0x^3 + a_1x^2 + a_2x + a_3 = 0, \qquad g \equiv b_0x^3 + b_1x^2 + b_2x + b_3 = 0$$

have a common root $x \neq 0$. Then

$$-b_0f + a_0g = (a_0b_1)x^2 + (a_0b_2)x + (a_0b_3),$$
$$(b_3f - a_3g)/x = (a_0b_3)x^2 + (a_1b_3)x + (a_2b_3).$$

By Ex. 1 of §112, the resultant of these two quadratic functions is

$$F = \begin{vmatrix} (a_0b_3) & (a_0b_1) \\ (a_2b_3) & (a_0b_3) \end{vmatrix}^2 - \begin{vmatrix} (a_0b_3) & (a_0b_1) \\ (a_1b_3) & (a_0b_2) \end{vmatrix} \cdot \begin{vmatrix} (a_1b_3) & (a_0b_2) \\ (a_2b_3) & (a_0b_3) \end{vmatrix}.$$

This is, however, not the resultant R of the cubic functions f, g. To show that (a_0b_3) is an extraneous factor, note that the terms of F not having this factor explicitly are

$$(a_0b_1)(a_2b_3)\{(a_0b_1)(a_2b_3) - (a_0b_2)(a_1b_3)\}.$$

The quantity in brackets is equal to $-(a_0b_3)(a_1b_2)$, since, as in Ex. 2 of §101,

$$0 = \tfrac{1}{2}\begin{vmatrix} a_0 & a_1 & a_2 & a_3 \\ b_0 & b_1 & b_2 & b_3 \\ a_0 & a_1 & a_2 & a_3 \\ b_0 & b_1 & b_2 & b_3 \end{vmatrix} = (a_0b_1)(a_2b_3) - (a_0b_2)(a_1b_3) + (a_0b_3)(a_1b_2).$$

We now see that $F = (a_0b_3)R$, where R is given in Ex. 2 of §112. This method of elimination therefore introduces an extraneous factor (a_0b_3). The student should employ only methods of elimination (such as those due to Sylvester, Euler, and Bézout) which have been proved to lead to the true resultant.

EXERCISES

Find the result of eliminating x and hence find all sets of common solutions of

1. $x^2 - y^2 = 9$, $xy = 5y$.

2. $x^2 + y^2 = 25$, $x^2 + 3(c-1)x + c(y^2 - 25) = 0$.

3. When $x^2 + ax + b = 0$ has a double root, what 3-rowed determinant is zero?

4. Find the roots of $x^6 + 3x^4 + 32x^3 + 67x^2 + 32x + 65 = 0$ by §79.

115. Discriminants. Let $\alpha_1, \ldots, \alpha_m$ be the roots of

$$(9) \qquad\qquad f(x) \equiv a_0x^m + a_1x^{m-1} + \cdots + a_m = 0 \qquad (a_0 \neq 0),$$

so that

$$(10) \qquad\qquad f(x) \equiv a_0(x - \alpha_1)(x - \alpha_2)\cdots(x - \alpha_m).$$

As in §44, we define the discriminant of (9) to be

$$D = a_0^{2m-2}(\alpha_1 - \alpha_2)^2(\alpha_1 - \alpha_3)^2 \cdots (\alpha_1 - \alpha_m)^2(\alpha_2 - \alpha_3)^2 \cdots (\alpha_{m-1} - \alpha_m)^2.$$

Evidently D is unaltered by the interchange of any two roots. Since the degree in any root is $2(m-1)$, the symmetric function D is equal to a polynomial in $a_0, \ldots, a_m$. Indeed, a_0^{2m-2} is the lowest power of a_0 sufficient to cancel the denominators introduced by replacing $\Sigma\alpha_1$ by $-a_1/a_0, \ldots, \alpha_1\alpha_2 \cdots \alpha_m$ by $\pm a_m/a_0$. By differentiating (10), we see that

$$f'(\alpha_1) = a_0(\alpha_1 - \alpha_2)(\alpha_1 - \alpha_3) \cdots (\alpha_1 - \alpha_m),$$
$$f'(\alpha_2) = a_0(\alpha_2 - \alpha_1)(\alpha_2 - \alpha_3) \cdots (\alpha_2 - \alpha_m),$$
$$f'(\alpha_3) = a_0(\alpha_3 - \alpha_1)(\alpha_3 - \alpha_2)(\alpha_3 - \alpha_4) \cdots (\alpha_3 - \alpha_m),$$

etc. Hence

$$a_0^{m-1} f'(\alpha_1) \cdots f'(\alpha_m) = a_0^{2m-1}(-1)^{1+2+\cdots+m-1}(\alpha_1 - \alpha_2)^2 \cdots (\alpha_{m-1} - \alpha_m)^2$$
$$= (-1)^{\frac{m(m-1)}{2}} a_0 D.$$

By (2), the left member is the resultant of $f(x)$, $f'(x)$. Hence

(11) $$D = (-1)^{\frac{m(m-1)}{2}} \frac{1}{a_0} R(f, f').$$

EXERCISES

1. Show that the discriminant of $f \equiv y^3 + py + q = 0$ is $-4p^3 - 27q^2$ by evaluating the determinant of order five for $R(f, f')$.

2. Prove that the discriminant of the product of two functions is equal to the product of their discriminants multiplied by the square of their resultant. Hint: use the expressions in terms of the differences of the roots.

3. For $a_0 = 1$, show that the discriminant is equal to

$$\begin{vmatrix} 1 & \alpha_1 & \alpha_1^2 & \cdots & \alpha_1^{m-1} \\ 1 & \alpha_2 & \alpha_2^2 & \cdots & \alpha_2^{m-1} \\ \hdotsfor{5} \\ 1 & \alpha_m & \alpha_m^2 & \cdots & \alpha_m^{m-1} \end{vmatrix}^2 = \begin{vmatrix} s_0 & s_1 & s_2 & \cdots & s_{m-1} \\ s_1 & s_2 & s_3 & \cdots & s_m \\ \hdotsfor{5} \\ s_{m-1} & s_m & s_{m+1} & \cdots & s_{2m-2} \end{vmatrix}$$

where $s_i = \alpha_1^i + \cdots + \alpha_m^i$. See Ex. 4, §88; Ex. 2, §102.

4. Hence verify that the discriminant of $x^3 + px + q = 0$ is equal to

$$\begin{vmatrix} 3 & 0 & -2p \\ 0 & -2p & -3q \\ -2p & -3q & 2p^2 \end{vmatrix} = -4p^3 - 27q^2.$$

5. By means of Ex. 1, §113, show that the discriminant of $a_0x^3 + a_1x^2 + a_2x + a_3 = 0$ is

$$- \begin{vmatrix} 2a_0a_2 & a_1a_2 + 3a_0a_3 & 2a_1a_3 \\ a_1 & 2a_2 & 3a_3 \\ 3a_0 & 2a_1 & a_2 \end{vmatrix} = 18a_0a_1a_2a_3 - 4a_0a_2^3 - 4a_1^3a_3 + a_1^2a_2^2 - 27a_0^2a_3^2.$$

MISCELLANEOUS EXERCISES

1. Find the equation whose roots are the abscissas of the points of intersection of two general conics.

2. Find a necessary and sufficient condition that

$$f(x) \equiv x^4 + px^3 + qx^2 + rx + s = 0$$

shall have one root the negative of another root. When this condition is satisfied, what are the quadratic factors of $f(x)$? Apply to Ex. 4, §74. Hint: add and subtract $f(x)$ and $f(-x)$.

3. Solve $f(x) \equiv x^4 - 6x^3 + 13x^2 - 14x + 6 = 0$, given that two roots α and β are such that $2\alpha + \beta = 5$. Hint: $f(x)$ and $f(5 - 2x)$ have a common factor.

4. Solve $x^3 + px + q = 0$ by eliminating x between it and $x^2 + vx + w = y$ by the greatest common divisor process, and choosing v and w so that in the resulting cubic equation for y the coefficients of y and y^2 are zero. The next to the last step of the elimination gives x as a rational function of y. (Tschirnhausen, *Acta Erudit.*, Lipsiae, II, 1683, p. 204.)

5. Find the preceding y-cubic as follows. Multiply $x^2 + vx + w = y$ by x and replace x^3 by $-px - q$; then multiply the resulting quadratic equation in x by x and replace x^3 by its value. The determinant of the coefficients of x^2, x, 1 must vanish.

6. Eliminate y between $y^3 = v$, $x = ry + sy^2$, and get

$$x^3 - 3rsvx - (r^3v + s^3v^2) = 0.$$

Take $s = 1$ and choose r and v so that this equation shall be identical with $x^3 + px + q = 0$, and hence solve the latter. (Euler, 1764.)

7. Eliminate y between $y^3 = v$, $x = f + ey + y^2$ and get

$$\begin{vmatrix} 1 & e & f - x \\ e & f - x & v \\ f - x & v & ev \end{vmatrix} = 0.$$

This cubic equation in x may be identified with the general cubic equation by choice of e, f, v. Hence solve the latter.

8. Determine r, s and v so that the resultant of

$$y^3 = v, \qquad y = \frac{x+r}{y+s}$$

shall be identical with $x^3 + px + q = 0$. (Bézout, 1762.)

9. Show that the reduction of a cubic equation in x to the form $y^3 = v$ by the substitution

$$x = \frac{r + sy}{1 + y}$$

is not essentially different from the method of Ex. 7. [Multiply the numerator and denominator of x by $1 - y + y^2$.]

10. Prove that the equation whose roots are the $n(n-1)$ differences $x_j - x_k$ of the roots of $f(x) = 0$ may be obtained by eliminating x between the latter and $f(x+y) = 0$ and deleting from the eliminant the factor y^n (arising from $y = x_j - x_j = 0$). The equation free of this factor may be obtained by eliminating x between $f(x) = 0$ and

$$\{f(x+y) - f(x)\}/y = f'(x) + f''(x)\frac{y}{1\cdot 2} + \cdots + f^{(n)}(x)\frac{y^{n-1}}{1\cdot 2 \cdots n} = 0.$$

This eliminant involves only even powers of y, so that if we set $y^2 = z$ we obtain an equation in z having as its roots the squares of the differences of the roots of $f(x) = 0$. (Lagrange *Résolution des équations*, 1798, §8.)

11. Compute by Ex. 10 the z-equation when $f(x) = x^3 + px + q$.

APPENDIX

THE FUNDAMENTAL THEOREM OF ALGEBRA

THEOREM.   *An equation of degree n with any complex coefficients*

$$f(z) \equiv z^n + a_1 z^{n-1} + \cdots + a_n = 0$$

has a complex (real or imaginary) root.

Write $z = x + iy$ where x and y are real, and similarly $a_1 = c_1 + id_1$, etc. By means of the binomial theorem, we may express any power of z in the form $X + iY$. Hence

(1) $$f(z) = \phi(x, y) + i\psi(x, y),$$

where ϕ and ψ are polynomials with real coefficients.

The first proof of the fundamental theorem was given by Gauss in 1799 and simplified by him in 1849. This simplified proof consists in showing that the two curves represented by $\phi(x, y) = 0$ and $\psi(x, y) = 0$ have at least one point (x_1, y_1) in common, so that $z_1 = x_1 + iy_1$ is a root of $f(z) = 0$. This proof is given in Chapter V of the author's *Elementary Theory of Equations*.

We here give a shorter proof, the initial idea of which was suggested, but not fully developed, by Cauchy.[1]

LEMMA 1. $a_1 h + a_2 h^2 + \cdots + a_n h^n$ *is less in absolute value than any assigned positive number p for all complex values of h sufficiently small in absolute value.*

The proof differs from that of the auxiliary theorem in §62 only in reading "in absolute value" for "numerically."

We shall employ the notation $|z|$ for the absolute value $+\sqrt{x^2 + y^2}$ of $z = x + iy$.

---

[1] For a history of the fundamental theorem, see *Encyclopédie des sciences mathématiques*, tome I, vol. II, pp. 189–205.

LEMMA 2.   *Given any positive number P, we can find a positive number R such that $|f(z)| > P$ if $|z| \geqq R$.*

The proof is analogous to that in §64. We have

$$f(z) = z^n(1 + D), \qquad D \equiv a_1\left(\frac{1}{z}\right) + \cdots + a_n\left(\frac{1}{z}\right)^n.$$

Since (Ex. 5, §8) the absolute value of a sum of two complex numbers is equal to or greater than the difference of their absolute values, we have

$$|f(z)| \geqq |z|^n[1 - |D|].$$

Let p be any assigned positive number < 1. Applying Lemma 1 with h replaced by $1/z$, we see that $|D| < p$ if $|1/z|$ is sufficiently small, i.e., if $\rho \equiv |z|$ is sufficiently large. Then

$$|f(z)| > \rho^n(1 - p) \geqq P$$

if $\rho^n \geqq P/(1 - p)$, which is true if

$$\rho \geqq \sqrt[n]{\frac{P}{1 - p}} \equiv R.$$

This proves Lemma 2.

LEMMA 3.   *Given a complex number a such that $f(a) \neq 0$, we can find a complex number z for which $|f(z)| < |f(a)|$.*

Write $z = a + h$. By Taylor's theorem (8) of §56,

$$f(a + h) = f(a) + f'(a)h + \cdots + f^{(r)}(a) \cdot \frac{h^r}{r!} + \cdots + f^{(n)}(a) \cdot \frac{h^n}{n!}.$$

Not all of the values $f'(a)$, $f''(a), \ldots$ are zero since $f^{(n)}(a) = n!$. Let $f^{(r)}(a)$ be the first one of these values which is not zero. Then

$$\frac{f(a + h)}{f(a)} = 1 + \frac{f^{(r)}(a)}{f(a)} \cdot \frac{h^r}{r!} + \cdots + \frac{f^{(n)}(a)}{f(a)} \cdot \frac{h^n}{n!}.$$

Writing the second member in the simpler notation

$$g(h) \equiv 1 + bh^r + ch^{r+1} + \cdots + lh^n, \qquad b \neq 0,$$

we shall prove that a complex value of h may be found such that $|g(h)| < 1$. Then the absolute value of $f(z)/f(a)$ will be < 1 and Lemma 3 proved. To find such a value of h, write h and b in their trigonometric forms (§4)

$$h = \rho(\cos\theta + i\sin\theta), \qquad b = |b|(\cos\beta + i\sin\beta).$$

Then by §5, §7,

$$bh^r = |b|\rho^r \{\cos(\beta + r\theta) + i\sin(\beta + r\theta)\}.$$

Since h is at our choice, ρ and angle θ are at our choice. We choose θ so that $b + r\theta = 180°$. Then the quantity in brackets reduces to -1, whence

$$g(h) = (1 - |b|\rho^r) + h^r(ch + \cdots + lh^{n-r}).$$

By Lemma 1, we may choose ρ so small that

$$|ch + \cdots + lh^{n-r}| < |b|.$$

By taking ρ still smaller if necessary, we may assume at the same time that $|b|\rho^r < 1$. Then

$$|g(h)| < (1 - |b|\rho^r) + \rho^r|b|, \qquad |g(h)| < 1.$$

Minimum Value of a Continuous Function. Let $F(x)$ be any polynomial with real coefficients. Among the real values of x for which $2 \leqq x \leqq 3$, there is at least one value x_1 for which $F(x)$ takes its minimum value $F(x_1)$, i.e., for which $F(x_1) \leqq F(x)$ for all real values of x such that $2 \leqq x \leqq 3$. This becomes intuitive geometrically. The portion of the graph of $y = F(x)$ which extends from its point with the abscissa 2 to its point with the abscissa 3 either has a lowest point or else has several equally low points, each lower than all the remaining points. The arithmetic proof depends upon the fact that $F(x)$ is continuous for each x between 2 and 3 inclusive (§62). The proof is rather delicate and is omitted since the theorem for functions of one variable x is mentioned here only by way of introduction to our case of functions of two variables.

We are interested in the analogous question for

$$G(x, y) = \phi^2(x, y) + \psi^2(x, y),$$

which, by (1), is the square of $|f(z)|$. As in the elements of solid analytic geometry, consider the surface represented by $Z = G(x, y)$ and the right circular cylinder $x^2 + y^2 = R^2$. Of the points on the first surface and on or within their curve of intersection there is a lowest point or there are several equally low lowest points, possibly an infinite number of them. Expressed arithmetically, among all the pairs of real numbers x, y for which $x^2 + y^2 \leqq R^2$, there is[2] at least one pair x_1, y_1 for which the polynomial $G(x, y)$ takes a minimum value $G(x_1, y_1)$, i.e., for which $G(x_1, y_1) \leqq G(x, y)$ for all pairs of real numbers x, y for which $x^2 + y^2 \leqq R^2$.

---

[2]Harkness and Morley, *Introduction to the Theory of Analytic Functions*, p. 79, prove that a real function of two variables which is continuous throughout a closed region has a minimum value at some point of the region.

Proof of the Fundamental Theorem. Let z' denote any complex number for which $f(z') \neq 0$. Let P denote any positive number exceeding $|f(z')|$. Determine R as in Lemma 2. In it the condition $|z| \geq R$ may be interpreted geometrically to imply that the point (x, y) representing $z = x + iy$ is outside or on the circle C having the equation $x^2 + y^2 = R^2$. Lemma 2 thus states that, if z is represented by any point outside or on the circle C, then $|f(z)| > P$. In other words, if $|f(z)| \leq P$, the point representing z is inside circle C. In particular, the point representing z' is inside circle C.

In view of the preceding section on minimum value, we have

$$G(x_1, y_1) \leq G(x, y)$$

for all pairs of real numbers x, y for which $x^2 + y^2 \leq R^2$, where x_1, y_1 is one such pair. Write z_1 for $x_1 + iy_1$. Since $|f(z)|^2 = G(x, y)$, we have

$$|f(z_1)| \leq |f(z)|$$

for all z's represented by points on or within circle C. Since z' is represented by such a point,

(2) $$|f(z_1)| \leq |f(z')| < P.$$

This number z_1 is a root of $f(z) = 0$. For, if $f(z_1) \neq 0$, Lemma 3 shows that there would exist a complex number z for which

(3) $$|f(z)| < |f(z_1)|.$$

Then $|f(z)| < P$ by (2), so that the point representing z is inside circle C, as shown above. By the statement preceding (2),

$$|f(z_1)| \leq |f(z)|.$$

But this contradicts (3). Hence the fundamental theorem is proved.

ANSWERS

Pages 2–3

1. $3i$.　　　**2.** 2.　　　**3.** $-20 + 20i$.　**4.** $-\frac{2}{3}$.　　　**5.** $(8 + 2\sqrt{3})$.

6. $\frac{1}{5}(6 + \sqrt{5}) + \frac{1}{5}(2\sqrt{5} - 3)i$.　　　**7.** $\dfrac{-9}{13} + \dfrac{19}{13}i$.

8. $\dfrac{a^2 - b^2}{a^2 + b^2} + \dfrac{2ab}{a^2 + b^2}i$.　　　**10.** Yes.

13. 3, 4 and -3, -4.　　　**14.** $\pm(5 + 6i)$.

15. $\pm(3 - 2i)$.　　　**16.** $\pm[c + d + (c - d)i]$.

Pages 6–7

2. -3, -3ω, $-3\omega^2$;　i, ωi;　$\omega^2 i$;　$R = \cos 40° + i \sin 40°$, ωR, $\omega^2 R$.

3. $\pm(1 + i)/\sqrt{2}$;　$\pm(1 - i)/\sqrt{2}$;　$\pm\omega^2$.

Page 9

4. -1, $\cos A + i \sin A$ $(A = 36°, 108°, 252°, 324°)$.

6. R^3, R^6, R^9.

Pages 10–11

5. $p(p - 1)$.　　　**6.** $(p - 1)(q - 1)(r - 1)$ if $n = pqr$.

Pages 15–16

1. 51.　　　**2.** 13.

Page 17

1. Rem. 11, quot. $x^2 + 5x + 8$.　　　**2.** -61, $2x^4 - 4x^3 + 7x^2 - 14x + 30$.

3. -0.050671, $x^2 + 6.09x + 10.5481$.

4. $x^2 - x - 6$, $x + 2$; $4, 3, -2$.

5. $x^2 - x - 6 = 0$, $3, -2$.

6. $2 \pm \sqrt{5}$.   **7.** $2x^2 - x + 2$.   **8.** $x^2 + 1$.

Page 19

1. $x^3 - 3x^2 + 2x = 0$.   **2.** $x^4 - 5x^2 + 4 = 0$.

3. $x^4 - 18x^2 + 81 = 0$.   **4.** $x^4 - 5x^3 + 9x^2 - 7x + 2 = 0$.

5. $b^2 = 4ac$.   **7.** By theorem in §18.

Pages 21–22

1. $x^3 - 6x^2 + 11x - 6 = 0$.   **2.** $x^4 - 8x^2 + 16 = 0$.

3. $1, 2$.   **5.** $4, \frac{3}{2}, -\frac{3}{2}$.   **6.** $1, 3, 5$.   **7.** $1, 1, 1, 3$.

8. $2, -6, 18$.   **9.** $-3, 1, 5$.   **10.** $5, 2, -1, -4$.

11. $y^2 - (p^2 - 2q)y + q^2 = 0$.   **12.** $y^2 - (p^3 - 3pq)y + q^3 = 0$.

13. (i) $y^2 - y(p^3 - 3pq)/q + q = 0$.
   (ii) $y^2 - q(p^2 - 2q)y + q^4 = 0$.
   (iii) $y^2 - (p + p/q)y + 2 + q + 1/q = 0$.

14. $p^3r = q^3$.   **15.** $2, 4, -6$.

Pages 22–23

1. $5, -1 \pm \sqrt{-3}$.   **2.** $1 \pm i$, $1 \pm \sqrt{2}$.

3. $x^3 - 7x^2 + 19x - 13 = 0$.   **4.** $4, 1 - \sqrt{-5}$, $x^3 - 6x^2 + 14x - 24 = 0$.

6. $\pm 1, 2 \pm \sqrt{3}$.   **7.** $\sqrt{3}, 2 \pm i$.

9. $x^3 - \frac{3}{2}x^2 - \frac{5}{4}x + \frac{7}{8} = 0$.   **10.** $2 + \sqrt{3}$, $x^2 + 2x + 2 = 0$.

11, 12. Not necessarily.   **13.** No.

Page 26

1. $19\frac{1}{4}$, 3.   **2.** 6.   **3.** 2.   **4.** 3.   **5.** $0, -7, -\frac{7}{3}$.

Page 28

1. $-1, -1, -6$.   **2.** $-2, 3, 4$.

3. 1, 3, 6.   **4.** $-2, -4$.   **5.** None.

Page 30

1. $2, -1, -4, 5$.   **2.** 9.

3. 8, 9.   **4.** $-12, -35$.   **5.** $2, 2, -3$.

Page 32

1. $1, 3, 9, \frac{1}{3}$.   **2.** $1, \frac{1}{2}, \frac{1}{3}$.   **3.** $-\frac{1}{6}$.   **4.** $\frac{1}{2}, -\frac{1}{4}, -\frac{1}{4}$.

5. $\frac{1}{4}, -\frac{1}{4}, \frac{1}{6}$.   **6.** $-\frac{1}{2}, \frac{1}{3}, \frac{1}{4}$.   **7.** $\frac{1}{2}$.   **8.** $\frac{2}{3}$.

10. $x^2 - 12x - 12 = 0$.   **11.** $x^3 - 3x^2 - 12x + 54 = 0$.

Page 34

1. 1, 4.   **2.** $-1, -4$.   **3.** $0.7, -5.7$.

4. $-0.7, 5.7$.   **5.** 2, 2.   **6.** Imaginary.

Pages 43–44

5. $x^5 + x^4 - 4x^3 - 3x^2 + 3x + 1 = 0$.

6. $-\frac{1}{2}(1 \pm \sqrt{-3})$, $\frac{1}{2}(7 \pm \sqrt{45})$.

10. See (11), §32.

11. Edges roots of $x^3 - 7x^2 + 12x - v = 0$, all real (§45) and irrational.

14. Δ = area, c = hypotenuse, squares of legs $\frac{1}{2}(c^2 \pm \sqrt{c^4 - 16\Delta^2})$.

15. Δ area, a, b given sides, square third side is $a^2 + b^2 \pm 2\sqrt{a^2 b^2 - 4\Delta^2}$.

16. $y^4 - 2y^3 + (2 - g^2)y^2 - 2y + 1 = 0$, pos. roots 0.09125, 10.95862.

Pages 48–49

3. $g = 2$, $R + R^8 + R^{12} + R^5$, etc., $z^3 + z^2 - 4z + 1 = 0$.

4. $g = 2$, $R + R^8$, $R^2 + R^7$, $R^4 + R^5$.

5. $\frac{1}{2}(1 \pm \sqrt{-3})$, $\frac{1}{2}(-5 \pm \sqrt{21})$.

6. $-1, 2 \pm \sqrt{3}, \frac{1}{2} \pm \frac{1}{2}\sqrt{-3}$.

7. $1, 1, 1, -1, \frac{1}{4}(1 \pm \sqrt{-15})$.

8. $-1, -2, -\frac{1}{2}, \frac{1}{6}(-5 \pm \sqrt{-11})$.

Pages 52–53

1. $-5, \frac{1}{2}(5 \pm \sqrt{-3})$.

2. $-6, \pm\sqrt{-3}$.

3. $-2, 1 \pm i$.

4. $\frac{1}{4}, \frac{1}{7}(-2 \pm \sqrt{-3})$.

Page 54

1. $\Delta = -400$, one.

2. $\Delta = 4 \cdot 27 \cdot 121$, three.

3. $\Delta = 0$, two.

4. $\Delta = 0$, two.

Page 55

1. $-4, 2 \pm \sqrt{3}$.

2. See Ex. 1, §47.

3. 1.3569, 1.6920, −3.0489.   **4.** −1.201639, 1.330058, −3.128419.

5. 1.24698, −1.80194, −0.44504.   **6.** 1.1642, −1.7729, −3.3914.

Page 57

1. 1, −1, $4 \pm \sqrt{6}$.   **2.** −1, −2, 2, 3.   **3.** $1 \pm i$, $-1 \pm \sqrt{2}$.

4. $1 \pm \sqrt{2}$, $-1 \pm \sqrt{-2}$.   **5.** 4, −2, $-1 \pm i$.

Page 61

1. $(-3, 9)$.   **2.** $\Delta = -250000$, $x = 3, -2, \pm i$.

3. $(3, 9), (-2, 4)$.   **4.** $h = 3$.   **5.** 6.856, 7.

Pages 67–68

2. 2.1.

3. $(-0.845, 4.921), (-3.155, 11.079)$;   between −4 and −5.

4. 1.1, −1.3.

5. Between 0 and 1, 0 and −1, 2.5 and 3, −2.5 and −3.

9. $120(x^3 + x)$, $120x^2 - 42$.

Pages 69–70

1. 3.   **2.** 2, −2.   **3.** −1.

4. Double roots, 1, 3.   **5.** None.   **6.** 3, 3, −3, 6.

Page 72

3. Use Ex. 3, p. 62, abscissas −1, 3.   **4.** Use Ex. 2, p. 62.

6. $y = -15x - 7$, $X^3 - 15X + 23 = 0$.

Pages 73–74

1. One real.

2. $(\pm\sqrt{\frac{7}{3}}, 7 \mp \frac{14}{3}\sqrt{\frac{7}{3}})$, three real.

3. $(\pm\sqrt{\frac{2}{3}}, -1 \mp \frac{4}{3}\sqrt{\frac{2}{3}})$, three.

4. $(-2 \pm \sqrt{5}, 23 \mp 10\sqrt{5})$, one.

Pages 84–85

13. $y^5 + 2y^4 + 5y^3 + 3y^2 - 2y - 9 = 0$.

14. $y^3 + 15y^2 + 52y - 36 = 0$.

Page 88

1. One, between -2 and -3.

2. One, between 1 and 2.

Pages 89–90

1. $(-4, -3)$, $(-2, -1)$, $(1, 2)$.

2. $(-2, -1)$, $(0, 1)$.

3. $(-2, -1.5)$, $(-1.5, -1)$, $(3, 4)$.

4. $(-2, -1)$, $(0, 1)$.

5. $(-7, -6)$, $(1, 2)$.

6. $(0, 1)$, $(3, 4)$.

Pages 92–93

2. 1, 1, 1, 2.

3. 1, 1, -2, -2.

4. 1, 1, two imaginary.

Page 95

1. $(-2, -1)$, $(0, 1)$, $(1, 2)$.

2. $(-4, -3)$, $(-2, -1)$, $(1, 2)$.

Pages 100–101

1. Single, -2.46955.

2. -1.20164, 1.33006, -3.12842.

3. 1.24698, −1.80194, −0.44504.   **4.** ±2.1213203,   2.1231056, −6.1231056.

5. 3.45592, 21.43067.   **6.** 2.15443.

7. −1.7728656,   1.1642479, −3.3913823.

8.   3.0489173, −1.3568958, −1.6920215.

9. 2.24004099.   **10.** 1.997997997.

11. 1.094551482.   **12.** 2.059, −1.228.

13. 1.2261.   **14.** 0.6527 = reciprocal of $2\cos 40°$.

15. 0.9397.   **16.** 1.3500.   **17.** 2.7138, 3.3840.

18. 5.46%.   **19.** 5.57%.   **20.** 9.70%.

Page 105

1. 2.   **2.** 3.

Pages 106–108

1. 2.24004099.   **2.** 2.3593041.   **3.** 1.997998.

Pages 109–110

1. 132°20.7′.   **2.** 157°12′.   **3.** 4.8425364.   **4.** 3.1668771.

5, 7. 15°16$\frac{1}{2}$′, 85°56$\frac{1}{2}$′, 212°49′, 225°57′.

6. 72°17′.   **8.** 5°56$\frac{1}{2}$′, 25°18′. **9.** 2.5541949.   **10.** 1.85718.

Page 111

1. −1.04727 ± 1.13594i.   **2.** $-\frac{2}{7} \pm \frac{1}{7}\sqrt{3}i$.

3. $-1 \pm i.$ **4.** $1 \pm i, 1 \pm 2i.$ **5.** $2 \pm i, \pm 2i.$

Pages 111–113

1. $217°12'27.4'' = 3.790988$ radians.   **2.** $42°20'47\frac{1}{4}''$ doubled.

3. $133°33.8'.$ **4.** $108°36'14''.$ **5.** $21.468212.$

6. Angle at center $47°39'13''.$ **7.** $49°17'36.5''.$

8. $1.4303\pi, 2.4590\pi, 3.4709\pi;$ $257°27'12.225''$ more exact than first.

9. $x/\pi = 0.6625, 1.891, 2.930, 3.948, 4.959.$

10. (i) $0.327739, 0.339224, 1124.333037.$
(ii) $0.250279, 0.894609, 1.127839.$
[Set $x = 1 + y, y = 1/z$ and solve by trigonometry.]

11. $3.597285.$ **12.** $10, 1.371288.$

13. $0.326878, 12.267305.$ **14.** $324°16'29.55''.$

15. 10 yr. 4 mo. 0 days.   **16.** $6.074674.$ **17.** $6.13\%.$

Page 116

1. $x = 5, y = 6.$ **2.** $x = 2, y = 1.$ **3.** $x = a, y = 0.$

Page 120

1. $-a_2b_1c_3d_4 + a_2b_1c_4d_3 + a_2b_3c_1d_4 - a_2b_3c_4d_1 - a_2b_4c_1d_3 + a_2b_4c_3d_1.$

2. $+, +.$

Page 126

3. $-3.$ **4.** $-8.$

Pages 129–130

1. $x = -8$, $y = -7$, $z = 26$.   **2.** $x = 3$, $y = -5$, $z = 2$.

3. $x = 6$, $y = 3$, $z = 12$.   **4.** $x = 5$, $y = 4$, $z = 3$.

5. $x = -5$, $y = 3$, $z = 2$, $w = 1$.   **6.** $x = 1$, $y = z = 0$, $w = -1$.

Page 133

1. Consistent: $y = -8/7 - 2x$, $z = 5/7$ (common line).

2. Inconsistent, case (β).   **3.** Inconsistent (two parallel planes).

4. Consistent (single plane).

5. (i) $z = -x - y - 2$.   (ii) inconsistent.   (iii) $x = \dfrac{a - 1}{a + 2}$, $y = z = \dfrac{-3}{a + 2}$.

6. (i) $x = \dfrac{(k - b)(c - k)}{(a - b)(c - a)}$.   (ii) $y = \dfrac{k - c}{a - c} - x$, $z = \dfrac{a - k}{a - c}$ if $k = a$ or $k = c$, but inconsistent if k is different from a and c. (iii) $z = 1 - x - y$ if $k = a$, inconsistent if $k \neq a$.

Page 134

1. $r = 2$, $x : y : z = -4 : 1 : 1$.   **2.** $r = 2$, $x : y : z = -10 : 8 : 7$.

3. $r = 1$, two unknowns arbitrary.   **4.** $r = 3$, $x : y : z : w = 6 : 3 : 12 : 1$.

5. $r = 2$, $z = -\frac{11}{3}x - \frac{19}{3}y$, $w = -\frac{10}{3}x - \frac{17}{3}y$.

Page 136

1. Ranks of A and B are 2; $y = -8/7 - 2x$, $z = 5/7$.

2. Consistent only when $a = -225/61$ and then $x = -\dfrac{5}{61}$, $y = \dfrac{3}{61}$, $z = \dfrac{45}{61}$.

3. Rank of A is 2, rank of B is 3, inconsistent.

4. A and B of rank 2, $x = 3$, $y = 2$.

Pages 140–142

1. $x = \dfrac{k(b-k)(c-k)(k+b+c)}{a(b-a)(c-a)(a+b+c)}$, if a, b, c are distinct and not zero and their sum $\neq 0$. If $a = b \neq c$, $ac \neq 0$, equations are inconsistent unless $k = 0$, a, c, or $-a-c$, and then $y = \dfrac{k(c-k)}{a(c-a)} - x$, $z = \dfrac{k(k-a)}{c(c-a)}$, x arbitrary.

3. $(a-b)(b-c)(c-a)$.   **4.** $(x-y)(y-z)(z-x)(xy+yz+zx)$.

6. $(a+b+c+d)(a+b-c-d)(a-b-c+d)(a-b+c-d)$.

7. $(a+b+c+d)(a-b+c-d)(a+bi-c-di)(a-bi-c+di)$.

11. $x_j = (k_1 - a_j) \cdots (k_n - a_j) \div \prod\limits_{\substack{s=1 \\ s \neq j}}^{n} (a_s - a_j)$.

12. $x(ab + ac + bc) = -abc$.

Pages 148–149

1. $\dfrac{p^4 - 3p^2 q + 5pr + q^2}{r - pq}$.

2. $\dfrac{(5p^2 - 12q)(p^2 - 4q)}{4(p^3 - 4pq + 8r)} - \dfrac{13}{4}p$.

5. $2p^2 - 2q$.

6. $24r - p^3$.

7. $\dfrac{3p^2 q^2 - 4p^3 r - 4q^3 - 2pqr - 9r^2}{(r - pq)^2}$.

8. $27r^2 - 9pqr + 2q^3 = 0$.

10. $y = q + r/x$.

11. $x = \dfrac{1 - py}{2 + 2y}$.

12. $y = \dfrac{4x^2 + px + q}{-3x - p}$, see §112.

13. $\dfrac{2q(p^3 + 2pq - r)}{p^2 q - pr + s} - 5p$, see Ex. 17.

Page 151

3. $s_2 = p^2 - 2q$,
$s_3 = p^3 - 3pq$,
$s_4 = p^4 - 4p^2 q + 2q^2$,
$s_5 = p^5 - 5p^3 q + 5pq^2$.

4. $s_{5n} = 5 \cdot 3^n$,
$s_k = 0$ if k is not divisible by 5.

5. All zero.

Page 155

2. See Ex. 2, p. 136.

3. $\epsilon^j \sqrt[5]{\frac{1}{2}c + \sqrt{Q}} + \epsilon^{5-j} \sqrt[5]{\frac{1}{2}c - \sqrt{Q}}, \quad Q = \frac{1}{4}c^2 - q^5$ $\qquad (j = 0, 1, 2, 3, 4).$

4. $\epsilon^j \sqrt[7]{\frac{1}{2}c + \sqrt{Q}} + \epsilon^{7-j} \sqrt[7]{\frac{1}{2}c - \sqrt{Q}}, \quad Q = \frac{1}{4}c^2 - q^7$ $\qquad (j = 0, 1, \ldots, 6).$

Page 156

1. $c_2^2 - 2c_1c_3 + 2c_4$.

2. $c_1^2 c_2 - 2c_2^2 - c_1 c_3 + 4c_4$.

3. $c_1 c_3 - 4c_4$.

4. $c_3^2 - 2c_2 c_4$.

Pages 157–158

1. $c_1 c_3 - 4c_4$ if $n > 3$, $c_1 c_3$ if $n = 3$.

2. $3c_1 c_4 - c_2 c_3 - 5c_5$.

3. $c_2 c_4 - 4c_1 c_5 + 9c_6$.

4. $c_3^2 - 2c_2 c_4 + 2c_1 c_5 - 2c_6$.

5. $y^3 - (p^2 - 2q)y^2 + (q^2 - 2pr)y - r^2 = 0.$

6. $y^3 - qy^2 + pry - r^2 = 0.$

7. $ry^3 + 2qy^2 + 4py + 8 = 0.$

8. Eliminate x by $y = s_2 - x^2$.

9. Use $p^2 - q + px = y$.

10. $-4 + pr/s$.

11. $(rs - pr^2 + 2pqs)/s^2$.

12. (i) $s_a s_b s_c s_d - \Sigma s_a s_b s_{c+d} + 2\Sigma s_a s_{b+c+d} + \Sigma s_{a+b} s_{c+d} - 6s_{a+b+c+d}.$
(ii) $\frac{1}{24}(s_a^4 - 6s_a^2 s_{2a} + 8s_a s_{3a} + 3s_{2a}^2 - 6s_{4a}).$

13. (i)

$$s_k = - \begin{vmatrix} 1 & 0 & 0 & \cdots & 0 & c_1 \\ c_1 & 1 & 0 & \cdots & 0 & 2c_2 \\ c_2 & c_1 & 1 & \cdots & 0 & 3c_3 \\ c_3 & c_2 & c_1 & \cdots & 0 & 4c_4 \\ \cdots\cdots\cdots\cdots\cdots\cdots\cdots\cdots \\ c_{k-1} & c_{k-2} & c_{k-3} & \cdots & c_1 & kc_k \end{vmatrix}, \quad s_3 = - \begin{vmatrix} 1 & 0 & c_1 \\ c_1 & 1 & 2c_2 \\ c_2 & c_1 & 3c_3 \end{vmatrix},$$

where all but the last term in the main diagonal is 1, and all terms above the diagonal are zero except those in the last column. If $k > n$, we must take $c_j = 0 \quad (j > n)$.

(ii)

$$k!\,c_k = -\begin{vmatrix} 1 & 0 & 0 & \cdots & 0 & s_1 \\ s_1 & 2 & 0 & \cdots & 0 & s_2 \\ s_2 & s_1 & 3 & \cdots & 0 & s_3 \\ \multicolumn{6}{c}{\dotfill} \\ s_{k-1} & s_{k-2} & s_{k-3} & \cdots & s_1 & s_k \end{vmatrix}, \quad 3!\,c_3 = -\begin{vmatrix} 1 & 0 & s_1 \\ s_1 & 2 & s_2 \\ s_2 & s_1 & s_3 \end{vmatrix}.$$

Page 167

1. $y^2(16 - y^2)$; $\quad y = 0$, $x = \pm 3$; $\quad y = \pm 4$, $x = +5$.

2. $(c-1)^2(y^2 - 25)(y^2 - 16)$. If $c \neq 1$, $y = \pm 5$, $x = 0$; $\quad y = \pm 4$, $x = +3$.

3. $\begin{vmatrix} 1 & a & b \\ 2 & a & 0 \\ 0 & 2 & a \end{vmatrix} = 4b - a^2.$ **4.** $2 \pm 3i$, $-2 \pm i$, $\pm i$.

Pages 169–170

2. $pqr - p^2 s - r^2 = 0$, $x^2 + r/p$, $x^2 + px + ps/r$.

3. $1, 3, 1 \pm i$.   **11.** See Ex. 15, p. 134.

INDEX

Numbers refer to pages.

www.ingramcontent.com/pod-product-compliance
Lightning Source LLC
Chambersburg PA
CBHW031957190326
41520CB00007B/275